*Makers of Modern Science*

# ENRICO FERMI:
# Pioneer of the Atomic Age

*Ted Gottfried*

Facts On File
New York • Oxford

**ENRICO FERMI : Pioneer of the Atomic Age**

Copyright © 1992 by Ted Gottfried

Facts On File, Inc.          Facts On File Limited
460 Park Avenue South        c/o Roundhouse Publishing Ltd.
New York NY 10016            P.O. Box 140
USA                          Oxford OX2 7SF
                             United Kingdom

**Library of Congress Cataloging-in-Publication Data**
Gottfried, Ted.
    Enrico Fermi / Ted Gottfried.
      p.  c.m. — (Makers of modern science)
    Includes bibliographical references and index.
    Summary: Describes the life and work of the Nobel prize-winning
  physicist known for his research in the area of nuclear energy.
    ISBN 0-8160-2623-8
    1. Fermi, Enrico, 1901–1954—Juvenile literature. 2. Nuclear
  energy—History—Juvenile literature. 3. Physicists—Italy—
  Biography—Juvenile literature. [1. Fermi, Enrico, 1901–1954.
  2. Physicists. 3. Nuclear energy.]  I. Title.  II. Series.
  QC16.F46G68  1992
  539.7'092—dc20
  [B]                                    92-14683

British CIP data available on request from Facts On File.

Facts On File books are available at special discounts when purchased in bulk quantities for businesses, associations, institutions or sales promotions. Please contact our Special Sales Department in New York at 212/683-2244 (dial 800/322-8755 except in NY, AK or HI) or in Oxford at 865/728399.

Text design by Ron Monteleone
Jacket design by Catherine Hyman
Composition by Facts On File, Inc.
Manufactured by R.R. Donnelley & Sons, Inc.
Printed in the United States of America

10 9 8 7 6 5 4 3 2 1

This book is printed on acid-free paper.

# CONTENTS

*With All My Love for*
*Benjamin & Elizabeth Coakley,*
*Kyle & Kourtney Gutierrez*
*—May Your Lives Be Filled with Peace*

# ACKNOWLEDGMENTS

I am indebted to the following people for their help on this book: my librarian wife Harriet Gottfried for ongoing critiquing, advice and patience; author Janet Bode for her YA expertise; Albert Machlin, P. E. and Peter Brooks for their technical assistance in simplifying complex concepts; the staffs of the Istituto Italiano di Cultura and the American Institute of Physics in New York City; and—as always—the personnel of various branches of the New York Public Library.

# 1

## THE BIG BANG
## 5:30 A.M., July 16, 1945

First came a blinding flash of light. Mountains 10 miles away were etched in sharp detail against the early morning sky. Next came a wave of intense heat as if the door to a giant oven had been flung open. A shock wave rippled over the desert forcing the official spectators flat against the ground.

**July 25, 1946—Expanding cloud from atomic blast rises 5,500 feet over Bikini Atoll, as photographed from a plane flying 8 miles away. Old warships are anchored nearby to measure shock radiation.** (New York Public Library)

The roar that followed was painful to their eardrums, a mighty thunder greater than any artillery barrage. It echoed and re-echoed, as a huge ball of fire shot up towards the heavens. This was followed by a giant mushroom cloud that rose to a height of 40,000 feet.

An atomic bomb had been exploded—the first in history—and the world would never be the same.

It was set off at 5:30 A.M. on July 16, 1945, from the top of a steel tower set up on the Alamogordo U.S. Army Air Base about 120 miles south of Albuquerque, New Mexico. Surrounding the tower were a variety of scientific monitoring devices. However, the explosion completely vaporized the tower as well as much of the recording equipment. It decimated the desert surface over an area with a circumference of 800 yards, melting the sand and then solidifying it into a hard, brittle, green material that looked and felt like glass.

Beyond this area, some 10,000 yards from where the base of the tower had been, concerned U.S. government dignitaries, military brass and scientists had observed the explosion from bunkers. They had worn protective dark glasses, but the flash of light had been so intense that they were forced to cover their eyes anyway. And then they had squeezed their hands against their ears to shut out the deafening sound.

One man among them did not notice the roar. He *really* did not notice it! His name was Enrico Fermi.

Fermi was the man who set off the chain reaction that made the bomb possible. When it came to his scientific work he was single-minded, and so he did not hear the explosion his colleagues would later refer to as "The Big Bang."

Scientific research is like sports. To score, the focus of the scientist must be narrow and intense to the exclusion of everything else around him. The batter never takes his eye off the ball, the hoopster shuts out everything but the court, the golfer always follows through—and the scientist focuses his complete attention on the task at hand and nothing else.

So it was that as the thunder of the bomb filled those around him with awe, Enrico Fermi continued to drop small pieces of paper from various heights (from over his head, from shoulder level, from waist-high, etc.) and to follow their path to the ground with a concentration that shut out everything else around him, including "The Big Bang."

He had started ripping sheets of paper into strips about 20 minutes before the bomb was detonated. Carefully, he had torn the strips into equal squares. Now, as he dropped them, they did not fall to his feet. They fell to the ground at varying distances from him, transported by the shock wave of the explosion.

Fermi carefully paced off these distances and recorded them. So deep was his involvement in this task that when his wife later asked him what the explosion of the world's first atomic bomb had been like, he could describe the blinding flash of light but had no recollection of the "strong, sustained, awesome roar which warned of doomsday" described by General Thomas Farrell, one of the military observers present. The time intervals at which the papers landed—measured in seconds and split seconds—were particularly significant and as far as Fermi knew then, there might never be another opportunity to repeat this simple experiment. So he dropped papers and he measured the distances at which they landed and would not let even the sound of the nuclear blast distract him.

From these measurements, Fermi was able to calculate the strength of the explosion and—to some degree—its ripple effect. With much of the measuring equipment at the base of the tower destroyed by the initial blast, his figures determined just how powerful the atom bomb really was. Later, during follow-up A-bomb tests, when more sophisticated and better protected precision recording devices were successfully used, it was shown that Fermi's bits of paper had provided highly accurate measurements. "Simple experiments," he would tell his students, "are always best."

Enrico Fermi was 43 years old that momentous morning in July 1945. He was a slender man of below average height, in good shape physically, with a body that was wiry but not athletically impressive. His face was aquiline, his jaw rounded, and his ears seemed slightly prominent because of a hairline that had been receding for at least 10 years. But his eyes were striking, deep and dark. They were eyes that could never know enough, the eyes of a Nobel Prize winner who had used that honor as a steppingstone to continue his work in an area he had pioneered: nuclear physics.

He was a family man with a wife, Laura, and two children. The oldest, age 14, was his daughter, Nella. Her brother, Giulio, was nine. Like Enrico Fermi himself, his wife and children had all been born in Italy and immigrated to the United States just before the outbreak of World War II. And like him, they had been classified as enemy aliens when the United States entered the war.

On December 8, 1941, the day after the Japanese attack on Pearl Harbor, President Franklin D. Roosevelt acted on the knowledge that Italy and Germany were allied with Japan in the Axis pact and declared that an "invasion, or predatory incursion is threatened upon the territory of the United States" by Italy and Germany as well as Japan, and therefore all Germans and Italian citizens on American soil were to be considered "alien enemies."

Technically, Fermi and his family were still Italians, and not Americans. They had arrived in New York City from Italy, via Sweden, in early January 1939—just one month short of three years before the United States declared war on Italy. Why had Fermi not become an American citizen during that period? The answer was simple—and frustrating. At that time the United States Immigration and Naturalization Law specified a five-year residency requirement before an application for citizenship could be filed. It would normally take another year or two for naturalization requirements to be met, tests to be passed, and citizenship granted.

In Fermi's case this period was slightly shortened. He and his wife Laura were sworn in as citizens of the United States on July 11, 1944—about five-and-a-half years after their arrival in this country. For two and a half of those years, Fermi had been classified an enemy alien at the same time that he played a key role in the work of the top secret Manhattan District Project—the official name of the project that developed the A-bomb.

His work required him to shuttle regularly between New York and Chicago. He and his family lived in New Jersey. Each time he had to apply for a travel permit as an enemy alien "to make a trip outside of his own community" and to carry on his person an official "endorsement of the United States Attorney" granting permission for the journey.

Trips had to be planned 10 days in advance to allow time for the paperwork to work its way through the bureaucracy. This was particularly galling to Fermi because the federal government had assigned the highest priority to the Manhattan Project and experiments were apt to take unexpected turns requiring prompt observation and quick decisions. Nevertheless, he had to travel by car, or train. He was not permitted to fly. President Roosevelt was firm that "no enemy alien shall undertake any air flight, or ascend into the air."

"A sense of irony," Fermi had remarked years earlier when confronted by what he considered to be the absurdities of the Italian dictator Mussolini's racial decrees, "is essential to the scientist." It stood him in good stead as one of an elite group granted top U.S. wartime security clearance, one privy to the most closely guarded military secret of the war, who at the same time had to—as one of his fellow scientists remarked—"raise his hand like a schoolboy asking permission to go to the bathroom in order to leave New Jersey."

Certainly Fermi was annoyed. At the same time, he had seen enough of totalitarianism in his native Italy to recognize that the strictures placed on him in the United States were quite minor by comparison. He might raise his eyebrows comically at the red tape, but his loyalty and patriotism to the country that had not yet accepted him was always steadfast.

We had to win the war. That was Fermi's conviction. The alternative—a Fascist victory—was too awful to consider. And victory meant developing an atomic bomb before the enemy did. But first it had to be proved that such a bomb was possible.

In theory the possibility was a logical conclusion of research begun by Fermi at the University of Rome in 1934. But many physicists held that Fermi's prior work in the field of nuclear physics was purely speculative and could have no practical application. Anything beyond theory, they said, was in the realm of science fiction.

Fermi's early war work at Columbia University in New York and at the University of Chicago was aimed at proving them wrong by doing what had never been done before. He planned and supervised the construction of a nuclear reactor to split the atom and set off the kind of chain reaction needed for a bomb to achieve maximum devastation. This process would unleash nuclear energy.

It was achieved on December 2, 1942, a year after Pearl Harbor, at an underground laboratory that had formerly been used as a squash court beneath the bleachers of the football stadium of the University of Chicago. Here the first atomic pile was erected in the shape of a doorknob. It was made from chunks of uranium interspersed with graphite bricks. In keeping with theories Fermi had spelled out many years before, the graphite was a retarding agent (like the brakes on a car in motion), creating a drag effect to slow down the highly explosive uranium. Fermi and his colleagues had created a pile in which—if his theories were correct—the chain reaction would begin spontaneously. They called the point at which it would begin the critical size, or critical mass. They were right about it setting itself off, but they misjudged the critical size.

One of the scientists present was Emilio Segre. Some months later, Fermi's wife Laura expressed concern to him over the dangers involved in Enrico's work. "Don't be afraid of becoming a widow," he told her. "If Enrico blows up, you'll blow up too."

Segre was doubtless remembering that day in Chicago when Fermi and his group overestimated how much of a pile should be created before the chain reaction would begin. They were still adding uranium and graphite when fission occurred, provoking the effect they were seeking. Immediately Fermi and the others found themselves struggling to control the overloaded, unstable and highly volatile pile. For long moments it was touch-and-go as to whether or not they could bring it under control. Finally they did. The pile stabilized and the fissioning uranium atoms split other atoms in a chain reaction.

If the scientists had not regained and maintained control over that reaction, as Segre later implied to Laura Fermi, her husband and his colleagues would not have been the only victims. A goodly portion of America's second most densely populated city at that time would have been laid waste. The dead would have been counted in the hundreds of thousands.

Dreadful as that risk was, it should be remembered that this was wartime. Not only that, the United States in 1942 was by no means winning the war. The Japanese had overrun the Philippines and a number of U.S. bases in the Pacific. The Germans and Italians controlled Europe, including a vast area in Russia. In North Africa, Generals Montgomery and Rommel were scoring see-saw victories against each other. On balance, globally, the Allies were still on the defensive.

Fermi knew that German scientists were also working on creating a nuclear device. He was convinced that whoever perfected such a weapon first would win the war. He knew that the future of the world would be determined by the winner of the race.

The Chicago chain reaction was a most important step towards that goal. By creating the reactor and demonstrating the splitting of the atom, Fermi took atomic power—and the bomb—out of the realm of theory. His accomplishment removed the "fiction" from the enterprise some of his peers had termed "science fiction."

Following the success of the Chicago experiment, Fermi, along with the other scientists, moved his family to Los Alamos, New Mexico, where work was beginning on the creation of the actual bomb. It was one thing to create a chain reaction in a nuclear reactor; it was another to create a shell to contain it that was both safe enough to transport and mechanically sensitive enough to be set off before impact. (Even ordinary bombs filled with conventional explosives that detonate on impact do not do nearly as much damage as bombs timed to explode before they strike the ground.) At Los Alamos, Fermi was chief of the Advanced Physics Department of the bomb production laboratory headed by Dr. J. Robert Oppenheimer. The practical experiments he supervised there led directly to the big bang. Along with his other work, they accounted for the tribute to him reported by the *New York Times*.

His fellow scientists, according to the *Times*, had designated Enrico Fermi "The Father of the Atomic Bomb."

## CHAPTER 1 NOTES

p. 3        "strong, sustained, awesome roar . . ." Laura Fermi, *Atoms in the Family: My Life with Enrico Fermi*, p. 239.

p. 4        "Simple experiments . . ." Emilio Segre, *Enrico Fermi, Physicist*, p. 169–170.

p. 4     "invasion, or predatory incursion . . ." *New York Times,* December 8, 1941, p. 1.

p. 4     "to make a trip outside of . . ." Laura Fermi, *Atoms in the Family: My Life with Enrico Fermi,* p. 168.

p. 5     "no enemy alien shall . . ." Laura Fermi, *Atoms in the Family: My Life with Enrico Fermi,* p. 168.

p. 5     "a sense of irony . . ." Pierre De Latil, *Enrico Fermi, The Man and His Theories,* p. 43.

p. 5     "raise his hand like a . . ." Pierre De Latil, *Enrico Fermi, The Man and His Theories,* p. 82.

p. 6     "Don't be afraid of becoming . . ." Laura Fermi, *Atoms in the Family: My Life with Enrico Fermi,* p. 237.

p. 7     "The Father of the Atomic Bomb . . ." *New York Times,* November 28, 1954, p. 25.

# 2

## A SPINNING TOP— 1901–1918

On September 29, 1901, in Rome, Italy, the third child in as many years was born to Alberto and Ida Fermi. The infant was a boy, and his parents named him Enrico. Soon after his birth he was sent away from his family and he did not again see his mother and father, nor his brother and sister, until he was two-and-a-half years old.

As had his brother Giulio, born the year before, Enrico was sent to the country to be looked after by nurses. Their mother, a former elementary school teacher whose maiden name was De Gattis, was too weak from having had her children so close in succession to look after them properly. Also, the family's apartment near the Rome railroad station had neither heat nor hot water. This was dangerous to the health of newborns.

It was intended that Enrico, like Giulio, would be gone only a few short months. However, he was sickly as both an infant and toddler, and so his return home was repeatedly delayed. When he finally did rejoin his family it was as a stranger.

Enrico was a painfully thin child, small for his age, far from sturdy, and repeated illness had left him cranky and quick to cry. Immediately he came into conflict with his mother who had carried over from her years as a schoolteacher a no-nonsense strictness. His sister Maria, the oldest, remembered an early clash between the two.

Reacting to the contrast between his easygoing nurse in the country and his stern mother, Enrico's small, dark face had twisted up and he had sniveled and then bawled. His mother had looked down at him from the heights of parenthood. "In this home, naughty boys are not tolerated!" she told him coolly.

Enrico stopped crying. What was the point? The authority, the power, was his mother's. He had none. He would not challenge; he would behave. The only battles worth fighting, after all, are those that can be won. It was the first instance of a practical attitude which would guide him through life. Nevertheless, there would be battles which were worth fighting, and he would fight them.

Surface conformity, patience and pragmatism were traits that came down to Enrico from his grandfather. Stefano Fermi, like his ancestors, was a peasant farmer in the fertile Po River Valley. He tilled the soil near Piacenza, but he did not own the land he farmed. It was owned by the duke who ruled the Duchy of Parma, one of the small states that made up the loose confederation of Italy during the last half of the 19th century. Stefano Fermi did not challenge the authority of the duke of Parma as did other peasants who joined in the rebellions for land reform during that period. Instead, he gave up farming and entered the duke's service, eventually rising to become country secretary, an influential position entitling him to a uniform with brass buttons etched with the name and emblem of his patron.

Enrico was only four years old when his grandfather, Stefano, died. In later years, his memory of him was hazy. He recalled him as gnarled and twisted like a question mark from arthritis. He remembered that the old man was puzzled that his grandchildren did not enjoy drinking wine as he himself had from childhood. And Enrico recollected him as kindly.

Enrico's own father, Alberto, had quite different memories of Stefano. Alberto was the second son among Stefano's many children. Stefano's attitude towards them was sink or swim. He forced Alberto to quit school and leave home with little to offer a harsh world in exchange for a livelihood.

Alberto, however, was a man of ambition and determination. He went to work for the growing network of Italian railroads as a clerk and began working his way up the administrative ladder. For more than 25 years he crisscrossed Italy working for the railroad. Finally his job took him to Rome.

**Giulio, Enrico (age four) and Maria Fermi.** (Author's collection)

Here he met and fell in love with the elementary school teacher Ida De Gattis. He was 41; Ida was 27. After a whirlwind courtship, they married. They had three children in rapid succession. The third was Enrico Fermi.

With his father often absent and his mother distant and showing an obvious preference for his brother Giulio over himself and his sister, Maria, young Enrico became withdrawn. He turned to mathematics and science to occupy his time. In the Rome apartment near the railroad station, still without heat, he would lose himself in books while sitting on his hands to keep them warm and using his tongue to turn the pages.

Only one person shared his interests and provided the warmth his parents didn't. This was his brother Giulio, his mother's favorite, an outgoing boy a year older than Enrico and very different from him. Giulio was bubbly, fun-loving, quick to laughter. Enrico was serious, less impulsive, more suspicious of people. However, through constant association many of Giulio's traits rubbed off on Enrico.

This was not apparent right away. Tragedy would make Enrico suppress these traits just as they were beginning to take hold. But they would come out later, when adolescence developed into young manhood.

During their pre-teens and teens, Enrico and Giulio shared a fascination with newly emerging technology. They designed and built electric motors and the motors worked. They drew plans for airplane engines. (In later years experts would look at these drawings and express disbelief that they could be the work of such young boys.)

This time spent with his brother was important to Enrico. By contrast, school was time merely to be endured. Enrico's teachers thought he lacked imagination and criticized his writing—both his scrawled handwriting and his style, which they judged crude. It did not help that Enrico was untidy and that his hair was often uncombed.

As far back as second grade this schoolteacher's son had failed to distinguish himself. On one occasion, his class was assigned to write a short essay about things that might be made of iron. Enrico's composition consisted of one sentence: "With iron one makes some beds." This was all he could think of, and the only reason he thought of it was that every day on his way home from school he passed a store sign which read "Factory of Iron Beds." His teacher took this as evidence of simplemindedness, and Enrico's mother tended to agree.

But school wasn't important to him. There were increasingly complex engines to be built. There were airplane motors to be designed. And there was his adored older brother Giulio to share the wonderful hours of creativity.

And then, suddenly, Giulio was gone.

It was the winter of 1915. The cold was bitter, the Fermi apartment, as always, unheated. Giulio developed a sore throat. Its cause was diagnosed as a simple infection—not very serious according to the doctor. But it got worse and it became difficult for 15-year-old Giulio to breathe.

An operation was recommended. The surgery would be minor, the doctor said. Giulio would be home the same day. His mother and sister could wait for him in the reception area while the operation was performed.

But something went wrong. Suddenly there were nurses telling Giulio's mother, "Don't worry; you should not worry." But worry about what? A simple surgical procedure? What could there be to worry about?

And then the surgeon was beside Ida Fermi, taking her hand, urging bravery, control. It was something about the anesthesia. It had not even been completely administered. Before it could be— the surgeon could not say why because he did not know— Giulio was dead.

Ida Fermi's despair over the death of her favorite son was so great that no one took much notice of Enrico's grief. She cried for hours on end. Or she sank into deep depressions during which she seemed to withdraw from life entirely.

Enrico too withdrew deep within himself. He made no friends and he distanced himself from his family. Rather than succumb to the pain of his loss, he concentrated on overcoming it. He forced himself to walk alone past the hospital where Giulio had died. He did this until he could look on the hospital with detachment.

Enrico was 14 years old now and he was deliberately compartmentalizing his life. Each compartment reflected a different Enrico. At home, with his family, he was more withdrawn and sullen than ever. But at school he had discovered that with very little effort he could excel, and he had moved easily from being a dull student to the top of his class. Also, he had become aware of his body, of the need and pleasure of exercise, and now there were hours each day between school and home which were spent playing ball, or French War (an Italian roughhouse game) with other boys his age. But it was the physical activity which was important to him and he never became close with these boys; he had no time for friends. His time was filled by his twin passions: mathematics and physics.

The boy lost himself in these subjects to block out Giulio's death. They led him to the Campo dei Fiori, a Roman outdoor market famed for its antiquities where secondhand books were sold cheaply. Here Enrico found a two-volume set dealing with mathematical physics written some 70 years previously by a Jesuit scholar named Father Andrea Carafa. The boy's enthusiasm for this work was so great that he broke his customary silence to share it with his sister Maria.

"You have no idea how interesting it is!" he exclaimed. "I am learning the propagation of all sorts of waves."

Maria smiled politely. She was interested in philosophy, religion and literature, but less so in science.

"It's wonderful! It explains the motion of the planets."

Another vague, polite smile.

"Fantastic!" Enrico had just finished the chapter on ocean tides.

Soon after this incident Enrico Fermi met Enrico Persico, a boy a year older than himself. Enrico Persico had been aware of Fermi and his interest in physics and mathematics for some time. He himself shared this fascination. But he sensed that Fermi put a wall between himself and other boys, a wall that had grown thicker since the death of his brother. Enrico Persico was reluctant to try to breach that wall.

One day, however, the two boys fell into a conversation about a science project each had worked on at different times in school. "Speculation" was the word Fermi used to describe the basis of their friendship. Indeed, they discovered on that very first occasion that the game of "What If—?" linked them in a quest for knowledge.

Soon they were setting science problems for themselves to solve jointly. They built, bought, borrowed or filched crude equipment for their experiments.

Fermi was the bolder, the more scientifically adventurous. Persico was more cautious, had his feet more firmly planted on the ground. By now Fermi was becoming more self-assured, even arrogant, and superficially outgoing, while Persico was introverted, self-effacing and reluctant to display his brilliance. If Fermi had no close friends aside from Persico, he was nevertheless accepted by his peers. Persico, who did not play sports, was even more of a loner than Fermi.

Both 15-year-olds, however, were fascinated by things they could observe but did not understand. They became intrigued by tops— perhaps the oldest toy there is—and the way they would right themselves no matter the angle at which they started to spin. Why did that happen? And as the top slowed down, what determined the point at which its axis slowly tilted so that it started to wobble? And when it wobbled, why did the crown of the top move in a circle which seemed at odds with the movement of the rest of the top?

Day after day, for hours on end, the two boys spun tops and observed them. They jotted down notes, worked on them, tore them up and started over again. They constructed elaborate theories, tested them, altered them, discarded them—and began anew.

The research they had conducted on their own had built on what they had learned in school and had even carried their knowledge to the university level. They were determined to figure things out for themselves.

Incredibly, they did. They did not know the technical word *precession*—which describes the first spin of the top—but they worked out the formula for it when they concluded that the top rights itself according to the ratio between the force of that first spin and the weight of the top. Neither did they know the term *nutation,* which defines the motion of the top at the wobbling point. But they nevertheless figured out that it was friction that determined the speed the spinning top must maintain before it starts to wobble, and that the circling motion of the crown of the wobbling top is caused by the loss of a specific amount of that friction.

*Precession. Nutation.* The young Enricos had rediscovered two theorems studied by graduate students. Also, they had innocently developed a working theory of the gyroscope that duplicated experiments performed by the French physicist Léon Foucault only 60-odd years earlier. *Gyroscope* (from the Greek) means "to view the turning." That, of course, was exactly what the two Enricos had done. And it was also what Léon Foucault had done. The boys had done it with a top, but the motion of that top, like that of Foucault's gyroscope, replicated the turning of the planet Earth itself.

Enrico Fermi first heard of Foucault when he described the results of the experiments with tops to Engineer Amidei. A colleague of Enrico's father, 37-year-old Amidei had made physics and mathematics his hobby for over 20 years. When he learned of Enrico's curiosity about these subjects, he took an interest in the boy.

"Is it true that there is a branch of geometry in which important geometric properties are found without making use of the notion of measure?" It was one of Enrico's first questions to Amidei and its complexity bowled the older man over. He was even more impressed when, after he replied that it was true, the boy asked: "But how can such properties be used in practice—for example by surveyors, or engineers?"

By way of an answer, Amidei lent him a book. As their acquaintance blossomed, he began to give Enrico mathematical problems—trick puzzles to tease him a bit, really—to solve. When he did so easily, Amidei came up with harder problems, and over the next few years he came to Enrico with problems that he could not himself solve.

Amidei recognized the boy's genius. When Enrico graduated high school, the older man persuaded his family to let the boy apply for a fellowship to the Reale Scuola Normale Superiore in Pisa. This was an exclusive institution that took only students of exceptional ability. There was a highly competitive admission examination and only those who scored among the highest were accepted.

Enrico Fermi, then 17 years old, took the examination on November 14, 1918. His assignment was to write an essay on characteristics of sound. He chose vibrating rods and strings as his focus, took off from an advanced theory known as the Fourier Analysis and actually extended the logic of the theory.

This paper, as was common, was read by the examiner assigned to interview Enrico. He was a professor of descriptive geometry visiting from the University of Rome. His name was Giuseppe Pittarelli, and at first he was skeptical.

"I found it almost impossible to believe that this was the work of someone as young as you," Professor Pittarelli told Enrico. "I want to satisfy myself by asking you, here and now, whether you are truly the author of the paper before me."

"Yes, sir. I am."

The professor asked more questions designed to test young Fermi's knowledge. Finally he was satisfied. "Young man," he said, "may I say that you are exceptional, truly exceptional." He told Fermi that he was sure to come in first in the competition for admission. And he concluded that even among the creme-de-la-creme who made up the student body of the lofty institution, he would prove—again—"exceptional."

The eminent professor could not have guessed that within a year young Enrico Fermi would be in danger of being expelled from the Scuola Normale Superiore because of his exploits as the school clown.

## CHAPTER 2 NOTES

p. 10       "In this home, naughty boys . . ." Robert Lichello, *Enrico Fermi: Father of the Atomic Bomb*, p 5.

p. 12       "With iron one makes . . ." Laura Fermi, *Atoms in the Family: My Life With Enrico Fermi, p. 16.*

p. 13       "Don't worry . . ." Laura Fermi, *Atoms in the Family: My Life With Enrico Fermi,* p. 16.

p. 14       "You have no idea . . ." Laura Fermi, *Atoms in the Family: My Life With Enrico Fermi,* p. 19 (Also Robert Luchello, *Enrico Fermi: Father of The Atomic Bomb,* p. 5).

p. 14       "Speculation . . ." and "What if—? . . ." Emilio Segre, *Enrico Fermi, Physicist,* Appendix I, *Letters to Enrico Persico,* p. 6–7.

p. 16       "It is true . . ." and "But how . . ." Emilio Segre, *Atoms in the Family: My Life With Enrico Fermi,* p. 8–9.

pp. 16–17   "Yes, sir. I am." and quoted material on this page drawn from: Robert Lichello, *Enrico Fermi, Physicist,* p. 6.

# 3

# WAR AND PEACE—
# 1902–1922

Italy, during Enrico Fermi's childhood, was a weak country and a poor one, under the thumb of its powerful neighbor, Austria. But it was strong compared to the crumbling Ottoman Empire of Turkey that ruled Libya, the North African land just across a narrow stretch of the Mediterranean Sea from Italy.

Early in 1911, when Enrico was nine years old, the French invaded Morocco, adjacent to Libya. Germany, which also had its eye on the copper and mineral deposits of the area, sailed a battleship to the southwestern coast of Morocco and landed troops there. Spain sent troops to support France.

War was avoided when a deal was worked out which granted France a "protectorate" over Morocco in exchange for giving Germany almost 100,000 square miles of the French Congo. The German foreign minister proclaimed that this would "have a calming effect," but it really meant that France, Spain, Germany and England were splitting up Africa—and its untapped riches—among themselves. Italy, however, was not about to be left out.

On September 28, 1911, Italy delivered an ultimatum to Turkey demanding that it turn over the Libyan port of Tripoli to the Italian navy. The Turks refused. Two days later war was declared on Turkey and Italy's warships blockaded Tripoli. An Italian battleship sank a Turkish torpedo boat in the harbor of Prevesa. The Italo-Turkish War had begun.

A number of European nations objected strenuously to Italy's rash action. But most Italians supported the war. In Rome crowds thronged the streets to celebrate small victories.

Enrico Fermi and his brother Giulio were caught up in the excitement. The two small boys stretched to wave their flags and were entranced by the shiny new weapons of war on parade. They were proud of their father who had a key role in moving troops and war supplies through Rome and south by railroad towards the front. Alberto's job was an important part of the war effort.

Not all Italians, however, supported the Libyan adventure. The most organized opposition came from the Italian Socialist Party. One of the party's more outspoken anti-war firebrands was a 28-year-old stonemason named Benito Mussolini.

Mussolini was the son of a village blacksmith who held strong Socialist and atheist views. His mother—like Enrico Fermi's mother—was a schoolteacher. Mussolini was named after the Mexican Indian revolutionary leader Benito Juárez and was raised in a left-wing environment.

In 1902, to avoid military service, Benito went to Switzerland. He worked as a stonemason, but his main activity was spreading Socialist doctrines. For this he was expelled from the country.

Back in Italy he was kept under police surveillance. Working on a Socialist newspaper in Trento, in 1908 he became committed to the Irredentist movement—a nationalist group whose aim was to get Italy out from under Austrian economic domination. He also was drawn to the harsh superman philosophy of the German Friedrich Nietzsche, which was used to justify beliefs in national and racial superiority and to discredit democracy.

Reading Nietzsche helped form Mussolini's view that "violence is the fundamental element of social transformation." By 1910 he was writing that "Socialism is war; and in war, woe to those who have humanitarian feelings."

But Mussolini did not consistently hold this view. He would change his ideas many times for personal advantage over the next 40-odd years. An early sign of this opportunism was his return to pacifism during the Italo-Turkish War. During two days in September 1911, when the war was not yet a month old, he led a workers' movement in Forli to oppose it. As a result, he was imprisoned five months.

"Italy needs class war, not imperialistic wars," Mussolini had insisted. "The Socialist aim should not be an enlarged colonialist Italy, but rather an Italy cultivated, rich and free."

(Twenty-five years later, Mussolini's air force would drop poison gas on Ethiopian civilians and his Fascist troops would seize Ethiopia's mineral-rich lands.)

But Mussolini was bucking popular opinion. When the Turks gave in and ceded Libya to Italy in October 1912, Rome exploded with the joy of victory. Again Enrico and Giulio Fermi joined the celebration, flushed with pride at their father's part in it.

They were patriotic Italians, who looked on the job their father had done with the railroad as a true act of patriotism. Yet the Libyans continued to fight the Italians in defiance of the peace treaty with the Turks and a much fiercer war was in the offing. This war would call into question Alberto Fermi's loyalty to his country.

When Alberto Fermi started working for the railroad in 1882, Italy had for some time been a nation of small states with a weak central government. Because conditions were unstable, rich Italians invested their money abroad. The Austrian government had seen this as an opportunity to exploit Italy and to turn it into a sort of colony without actually ruling it. Austrian bankers had been subsidized to develop Italian industry. Labor was cheap and the profits could be both reinvested and siphoned off to Vienna. Austrian government policy promoted investment in Italy.

One major industry slated for development was Italian railroads. The Austrians built many small rail lines throughout Italy. Austrian managers and overseers ran them.

The small railroad for which Alberto Fermi worked merged with another small line. Through further mergers, Alberto came to work for a large Austrian parent company. He rose rapidly through its management structure. Soon he was a key man in the interlocking railroad system which crisscrossed Italy.

This system was nationalized by the Italian government in 1905. However, since only the Austrians had the expertise to run the railroads, they were left in control. Austrian investments were honored. Railroad profits continued to be shared generously with Vienna.

Alberto continued to work with the Austrians. He respected them and they valued his management skills, as did the Italian government which awarded him the title of *cavaliere.* Subsequently he was appointed *capo divisione,* a civil rank on a level with that of brigadier general in the Army.

By 1912, however, there was a change in how Italians viewed the massive Austrian investment in their country. Where at first it had been welcomed because it created jobs, now it was resented as exploitation by foreigners. On the eve of World War I, this hatred of Austrians had grown to include those who worked with them—executives like Alberto Fermi.

Such feelings reached a boiling point in 1914 when Italy was still pledged by the Triple Alliance to fight alongside Austria and Germany. The conservative Italian government supported the treaty. With World War I on the horizon, a nationwide military draft was contemplated.

In June, a little more than a month before the war started, riots broke out at Ancona and a general strike was called to protest a draft and the government's pro-Austrian policy. It spread and sparked an uprising of peasants against landowners.

Mussolini and other Socialist leaders tried to turn this uprising into a national revolution. They succeeded in capturing many towns and even provinces and set up revolutionary councils to govern. In the end, though, the government was able to put down the revolt and order was restored.

But inevitably, war broke out on August 1, and two days later Italy backed out of the Triple Alliance and announced that it would remain neutral. During the next two months, Mussolini twisted like a pretzel.

As editor of the official Socialist Party newspaper, at first he threatened the government with another uprising if it sided with Austria and Germany. Only neutrality would satisfy him. But when neutrality was proclaimed, he was not satisfied.

Courted by the French government, Mussolini did an about-face. He came out for Italy entering the war *against* Austria. Though many Italians approved, the decision was against Socialist anti-war policy and Mussolini was expelled from the Socialist Party.

Meanwhile, Alberto Fermi and his family were beginning to be subjected to prejudice because of his Austrian business connections. Both Giulio and Enrico had to defend their father's patriotism to schoolmates. Even after Giulio's death, Enrico and Maria had to endure snide remarks about "Austrian-lovers."

Mussolini continued to stir up anti-Austrian sentiment. Denounced by the Socialists, he founded the Fasci d'Azione Rivoluzionaria—the first Fascist Party—and in its newspaper, *Il Popolo d'Italia,* he insisted that "more liberty would exist in Europe and the proletariat would have better opportunities to develop its class capacities" if Italy fought alongside France and England.

Both the government and the Socialists denounced Mussolini's activities. The government arrested him. But he was released and a short while later fought a duel with Socialist leader Claudio Treves. The pacifist Treves wounded the ex-pacifist Mussolini— but not seriously.

A month later, in May 1915, Italy declared war against Austria. As in the Italo-Turkish war, railroads were crucial to moving troops and supplies. But the traditional Austrian involvement in them made their every move suspect. Delays, overcrowding, re-routing, accidents—all were labeled sabotage by a public suspicious of railroad managers whom they felt had worked too long with the Austrians to be trusted. Every day Alberto Fermi encountered this distrust. However, his knowledge of railroading was such that the wartime administration had to keep him on the job.

This affected young Enrico, of course. With Giulio gone, he had already become increasingly withdrawn. Now he lost himself in physics and mathematics to counter the prejudice. For him it was once again a case of the only battles worth fighting are those that can be won.

Italy, however, was not winning battles against the Austrians. The Libyan war had left the army short of manpower and munitions. The loss of the Austrian managers hurt war production. Increasingly the army was dependent on the railroads to move scarce materials quickly.

A front had been established at Isonzo, but after a six-month campaign and 280,000 men lost, the Italian army had made no headway. The war dragged on with a constant pressure to supply troops fighting at Trentino and Trieste. By August of 1916, the people were sick of the war. Pope Benedict XV labeled it "useless carnage," while Socialist Claudio Treves vowed, "Next winter not a man in the trenches!"

Mussolini, however, was more staunch than ever. He adopted a bit of graffiti scrawled on a wall by soldiers during the first days of the war and made it his motto: "It is better to live a day as a lion than a year as a sheep." He repeated this constantly during the war, and later he made it a slogan of his Fascist Party.

By now he had been drafted, wounded and discharged. However, Mussolini's wound was not suffered in battle. It was caused by the explosion of a hand grenade during training. Once out of the army, he continued to be pro-war.

The losses of the Italian army at Caporetto, a major defeat, undermined his efforts. Only the Italian air force seemed capable

**1915–1918—Words written by soldiers on a wall along the Piave River during World War I, which were an inspiration to schoolboys like Enrico Fermi, later became a slogan of Mussolini's Fascist Party. Translation: "It is better to live a day as a lion than a year as a sheep."** (Italian Tourist Service)

of achieving glory in the war. Its exploits thrilled schoolboys who could identify planes and knew aviators by name. Enrico Fermi, always fascinated by aircraft, was no exception.

Caporetto also affected the boy. Heavy losses because of mistakes made by the generals were blamed on sabotage by Austrian sympathizers who allegedly kept reinforcements from reaching the front in time to turn the tide of battle. Alberto Fermi was never directly charged with this, but the general air of suspicion was most unpleasant for him and his family.

When the war ended, Enrico, 17, felt relieved. Some 600,000 Italian men had died. If the war had gone on, he would have been called up. Enrico would not have been able to attend the Scuola Normale Superiore.

In July 1922, after four years at the Scuola, he was awarded his degree as doctor of physics. During that time he had paid little attention to the changes which Italy was going through. These changes increasingly involved Mussolini.

**November 1922—Benito Mussolini is ceded control of the Italian government by King Vittorio Emanuele III.** (Istituto Italiano di Cultura)

As the Socialist Party became strong after the war, there were a series of strikes. Industrialists and landowners financed Mussolini and his new Fascist Party to preserve their interests. He would provide the manpower necessary to break the strikes and destroy the trade unions and the left-wing parties. His enemies now were the Socialists, the Communists and organized labor.

Between 1919 and 1922, he struck against them repeatedly and violently. In October 1922, armed Fascist squads—the famous Black Shirts—staged the "March on Rome." Neither the police nor the army attempted to interfere. Enrico Fermi, just back from Pisa, witnessed the show of strutting Fascist power.

Their demand was that the reins of government be handed over to Mussolini. At first King Vittorio Emanuele III refused. But with the parliamentary government in shambles and the army's loyalty in doubt, he finally gave in. On October 30, Mussolini arrived in Rome by sleeping car from Milan to take control of the government of Italy.

It was a momentous event, but politics was not a major interest of young Enrico Fermi. He was busy making plans for his future as a scientist. Those plans would place him on a collision course with the dictator whose iron fist would rule Italy for the next 20 years.

## CHAPTER 3 NOTES

p. 18     "protectorate" and "have a calming effect . . ." *Chronicle of the 20th Century,* p. 156.

p. 20     all quoted material on this page drawn from: *Encyclopaedia Britannica,* Volume 15, 1970, p. 1099.

p. 22     "more liberty would exist . . ." *Encyclopaedia Britannica,* Volume 15, 1970, p. 1099.

p. 23     "useless carnage . . ." *Encyclopedia Britannica,* Volume 12, 1970, p. 773.

p. 23     "it is better to live . . ." Photo translation, Istituto Italiano di Cultura.

# 4

## CLOWN AND GENIUS— 1918–1924

At age seventeen, Enrico Fermi was both the pride and despair of his teachers at the Scuola Normale Superiore in Pisa. They judged his work exceptional. But his expertise in science too often spilled over into pranks which made those same teachers furious. Their best student was also their most mischievous.

His partner in horseplay was a fellow student and friend, Franco Rasetti. One winter night in 1920, they turned a homework assignment in chemistry into the making of a stink bomb. They brought it to a class lecture the next day.

The classroom was very cold. Enrico complained about it in a whisper to Franco, who sat across the aisle from him. Snickering, Franco reminded him that "Napoleon founded this school and compared with Russia, he probably thought it was the Sahara."

"Silence!" The professor was annoyed by their whispering. "There must be no further interruptions!"

But when he turned his back, Fermi hissed to his friend that "the enemy has been sighted. Prepare to fire."

"Ready to fire, sir."

"Fire away!"

The stink bomb was exploded. Its evil-smelling fumes drove the students and professor from the classroom.

Fermi and Rasetti were hauled before a faculty disciplinary committee. The stink bomb was not their only offense. Evidence of ongoing mischief was presented. It came out that Fermi and Rasetti had formed an Anti-Neighbors Society dedicated to tomfoolery.

This included dousing passersby with water from rooftops, rigging pails of water above doors to drench fellow students and even instructors, and imprisoning students in their room by screwing metal eyelets in doors and door frames and then padlocking them together. (Fermi had done this to Rasetti who had been furious because Fermi had gathered a crowd of students to witness his frustration at being unable to get out of his room.)

Ordinary citizens of Pisa had also been victims of the Anti-Neighbors Society's deviltry. Fermi and Rasetti had discovered that small pieces of metallic sodium when tossed into water would catch fire and explode. They would pitch these into the freshwater fountains adjacent to the public lavatories of Pisa to startle the men using the urinals.

Another of their pranks was described by one of the faculty discipline committee members: "One boy engages an innocent person in conversation while the other slips the link of a padlock through both buttonholes of his coat and snaps it shut, thus imprisoning him until the poor victim can either saw through the lock and free himself, or tear his clothes."

The disciplinary committee was not amused. Both boys realized that they had gone too far. Their pranks no longer seemed quite so funny. The sentiment of the committee was obvious. It was that the troublesome pair should be expelled.

They would have been had not their physics teacher, Professor Luigi Puccianti, spoken up in their behalf. Stressing that their pranks were "not to be condoned," he pleaded that they were caused by high spirits and youth and went on to say that "I wish to emphasize the extraordinary scholastic achievements of these two young students. Perhaps," he suggested gently, "they are bored that we teachers cannot keep up with them."

He asked for leniency for the boys and the committee was swayed. Fermi and Rasetti were not expelled. Instead, they were placed on probation and allowed to continue their studies.

During the rest of his time at the Scuola Normale Superiore, Enrico followed a middle-road between the "high spirits" Professor Puccianti had identified and his first reaction to the school. On December 9, 1918, shortly after his arrival in Pisa, Enrico had written to his best friend Enrico Persico that "during the first days of the new life I was slightly despondent." This may have been because the 16th-century palace in which he was housed did not live up to its glittering facade. The students' rooms were little more than unheated cells. He had to study with a *scaldino* (an iron pot in which charcoal burned and ashes collected) in his lap to keep his hands and stomach warm. The food, mostly dried codfish, was tasteless and not filling.

To add to this, as Professor Puccianti would later suggest, Enrico was bored. Much of the work he was called on to do in his first term duplicated experiments he had previously done by himself. He had already gone far beyond the classwork.

At first he countered this boredom by continuing to read advanced treatises in physics on his own. He read Poincaré and Appell and Planck's *Thermodynamics.* He formulated complex problems for himself and spent hours solving them and then working out proofs for the solutions. He developed the habit he would have all his life of spending leisure time reviewing theorems of physics in his head.

Enrico came across scientific papers on the hydrogen atom by the Danish physicist Niels Bohr. In 1922 Bohr would win the Nobel Prize for his work on atomic structure, but at this time he was largely unknown in Italy even among scientists. Yet young Fermi immediately saw the implications of Bohr's work and this may have inspired his own career in nuclear science.

His scientific reading continued after he became friends with Franco Rasetti. Enrico was quite able to work in their monkey-shines along with his classwork and independent research. He could not be distracted from science when he did not want to be. Even so, his future wife would write about their misadventures, stating that Enrico was "dragged into it and held fast there by Franco Rasetti."

Franco was an unusual boy. Although his real interest was the natural sciences—botany, entomology, paleontology etc.—he chose to major in physics at university. He did this "because it was not easy for him to understand physics and he wanted to prove to himself that he could overcome any difficulty."

After their near-expulsion, both boys buckled down. In their spare time they went to the top of the famous Tower of Pisa and tried to envision Galileo performing his famous experiments on falling bodies from there. They also went to the adjacent cathedral to view the very lamp whose swaying had inspired Galileo's Laws of the Pendulum. On weekends they went bicycling, or climbed the mountains of the Alpi Apuane north of Pisa.

The exercise sharpened Enrico's mind for his studies. By now he had gone beyond his professors and was beginning to break new ground in theoretical physics. One day Professor Puccianti kept Enrico after class.

"I am an ass!" the professor announced abruptly. "But you are a lucid thinker and I can always understand what you explain." He went on to ask Enrico to teach him Einstein's theory of relativity.

Enrico, never a braggart, but never falsely modest either, agreed to teach his teacher. At this same time, Enrico was plugging up what he considered to be a hole in his scientific knowledge. He wrote his friend Enrico Persico about this on December 11, 1919.

"I have decided to study chemistry, theoretical and non-theoretical," he explained. "The lectures, for the time being, do not keep me very busy—eight hours a week. To these I add six hours a day of chemistry plus some physics lab . . ."

But Fermi, along with his friend Rasetti, cut corners on chemistry homework. Instead of going through a lengthy chemical analysis of powdered substances as they were supposed to, they put them under the microscope and identified the ingredients that way. They would then work up a detailed, point-by-point account of the "analysis" they hadn't done.

It wasn't that Fermi wanted to cheat. It was just that he could never stand doing something the hard way if an easier method was available. He had the kind of mind which always saw the easier way. It was a key ingredient of the genius that led to his splitting the atom.

Everyone at the school, faculty and students, was beginning to recognize that genius by January 1920. "At the physics department I am slowly becoming the most influential authority," he wrote Enrico Persico. "In fact, one of these days I shall hold, in front of several magnates, a lecture on the quantum theory, for which I am always a great propagandist."

The quantum theory had been formulated by Max Planck and refined by Albert Einstein in the early 1900s, but was still in question at this time. (Later the work that won Niels Bohr the Nobel Prize would further expand on it, as would experiments performed by Fermi himself.) Its initial concern was with the nature of light.

Looking at light, it seems to be a steady stream. But quantum theory questions what the eye sees. It says that a beam of light is not continuous but a series of many small units in action. It is like a child's cartoon flip-book. In the flip-book each page has a fixed drawing. The rapid flipping of the pages makes the stationary cartoon figures seem to move so that the eye "sees" action. So it is with light. We see a "beam," but it is really a series of tiny units, particles the scientists call *quanta,* or *photons,* in movement.

That is the original quantum theory that caught and held young Enrico Fermi's interest. There were already attempts at the beginning of the 1920s to extend it to areas other than light. If light was viewed as energy, after all, why should quantum theory not apply to other forms of energy?

For many years, Fermi's work would have to do with the possibilities this question presented. The answers would reshape the science of physics. But this was not Enrico's only interest during his last two years at the Scuola Normale.

He was also trying to expand on Einstein's theory of relativity. While still a student he wrote a paper called "On the Phenomena Occurring Near a World Line," which proved to be a valuable contribution to mathematical theory.

In July 1922, Fermi went before the university examining board that would decide if he should receive his doctoral degree. There were 11 examiners, all wearing square-topped hats and black togas. Fermi also wore the traditional black toga.

The examination was held in the Aula Magna, a large assembly room. Many of Fermi's friends came to hear his oral presentation. They expected that it would be a great success. His written doctoral thesis on X rays was already being discussed and praised.

But the results of Fermi's talk to the examining board were disappointing. He was simply over their heads. They fidgeted, and yawned, and pointedly took out their pocket watches and looked at them.

When he was done, none of the examiners shook hands with him, as was the custom. None congratulated him. He received his doctor of physics degree *magna cum laude* (with high honors), but the university did not publish his thesis, as was customary when such an honor was awarded.

Fermi had no time to brood over this treatment. Immediately after graduation he returned to Rome to meet with Professor Orso Mario Corbino, a prominent physicist who was also a member of the Italian Senate. Enrico wanted guidance in planning his future. The meeting took place on October 28 as Mussolini's Black Shirts were marching in the streets of Rome.

News of their demand that the king hand over the government to Mussolini had just reached Professor Corbino and he was very upset. If the king complied, he was sure that the Senate would be dissolved and—even more terrible—that there would be a civil war. He feared a massacre by the army, or the Fascist legions, or both. "However," he told young Fermi glumly, "if the King doesn't sign, we are certainly going to have a Fascist dictatorship under Mussolini."

Fermi was upset by this wise man's judgment. For a brief moment he focused on politics rather than science. He told his family that the situation could mean that young people like himself might have to emigrate.

"Emigrate?" His sister Maria had stared at him. "Where to?"

"Somewhere. . . . The world is large."

But Fermi did not emigrate—not then. What he did do, the following winter, was go to Göttingen, Germany, to study with one of the world's most illustrious physicists, Max Born. With the help of Professor Corbino, he had received a grant from the Italian Ministry of Public Instruction.

In Göttingen a new side of Fermi's character revealed itself. Germany, which had never recovered economically from World War I, was in the grip of dizzying inflation. The value of the German mark was plummeting. Only foreign money had any real value. Enrico was receiving a weekly payment as part of his fellowship from the Italian government. He became expert at selling his *lire* at a profit. He learned that he could double his profit just by waiting until evening when the price was at its highest. He lived better than he ever had and he bought himself a new bicycle.

Far more important to Fermi, however, was the chance to get deeper into quantum theory with Max Born. They did not develop a very warm relationship, but even so, Born helped Fermi understand how the structure of matter related to the quantum theory.

All matter is made up of tiny units called *atoms.* Each atom is a tiny solar system. The "sun" of this system, its center, is called the *nucleus.* This nucleus is made up of *protons* and *neutrons*—units of positive and negative energy. "Planets" called *electrons* revolve around this nucleus, this "sun." Each electron always follows the same path around the nucleus and that path is called the electron's *orbit.* As long as nothing disturbs the electron, it follows that orbit just as Earth and the other planets follow their orbits around our sun. If the orbit is not disturbed, the electron neither gives off nor takes in any energy.

But what would happen if a large meteor were to crash into the Earth? The Earth would take in energy from the force behind the blow and it would give off energy released by the impact. So it is with an electron. If something strikes it, then it both absorbs and releases energy. However, an electron is so small, that it can only be struck if the atom containing it is struck.

Picture a pea sealed inside a ping-pong ball. You can't smash the pea without smashing the ball. Likewise, you can't strike an electron without smashing an atom.

If some outside force does smash the atom, however, the electron is forced to change its orbit. Depending on the nature of the blow, it may be forced into a smaller or a larger orbit. If smaller, it will use less energy, and the leftover energy will be given off by the atom. If larger, then the energy will be absorbed by the atom.

These changes are called *radiation* and *absorption.* The units of energy released are called *quanta.* The amount of quanta which a substance can release following an impact depends on what kind of substance it is.

If the quantum theory were to look at matter as it looked at light—in other words as energy made up of individual cells in motion rather than as a solid, then those cells would be atoms. If an atom could somehow be split, then the force involved would release energy from the electrons circling the nucleus of the atom. This was the kind of pure theory which Fermi delighted in.

But it was only theory, of course. Back then even he did not foresee its having any practical application. There was as yet no machine capable of crushing an atom. That was several years in the future. And certainly there was no thought of a bomb.

The following year, 1923, while on leave, he pursued his studies while teaching a course in elementary mathematics at the University of Rome. He was back from Germany teaching in Rome on May 8, 1924, when his mother died.

Their relationship had always been difficult. Enrico had never been able to replace his brother Giulio in her affections. Mostly she was either strict with him or ignored him.

Yet his mother had passed on to him many of the qualities needed for success in science. Fed up with an old-fashioned oven which wouldn't cook as she wanted it to, for instance, Ida Fermi had designed and built a homemade pressure cooker years in advance of the manufacture of such labor-saving devices. "If she wanted something, she would make it for herself," Fermi would remember. Despite the distance between them, he was proud of her.

Not long after his mother's death, Enrico went on a picnic hike with some friends. On the streetcar which took them out of Rome to a rippling meadow where the Tiber and Aniene rivers meet, introductions were made among members of the group who had not met before. Fermi found himself shaking hands with a vivacious young girl of 16 named Laura Capon.

Enrico, who was 22, regarded her as a child. When the group broke up into teams to play soccer, he took her under his wing. "You shall be goalkeeper," he told her. "Just try to catch the ball when it comes through the goal. If you miss it, don't worry, we shall win the game for you."

Laura was not convinced of his assurances. He was grinning at her and she could clearly detect a baby tooth that made him look a little like a playful rabbit. His appearance was at odds with the fatherly attitude he was taking towards her. She suspected he might not prove as expert a soccer player as he implied.

She was right. Early in the game he tripped trying to kick the ball and sprawled face down on the ground. Laura was watching him, giggling, when a goal was attempted by the opposing team. The ball hit her in the chest, knocking the wind from her. But the accident cost the other side the goal and the Fermi team won the game.

Such was the first meeting between Enrico Fermi and Laura Capon. She thought him too old, inclined to brag and somewhat bumbling. He thought her a child whose unseemly giggles betrayed a lack of respect for her elders.

Not that it mattered. They did not see each other for two years after that first encounter. By then they had forgotten all about each other. And if they had remembered, they would not have anticipated that they would fall head-over-heels in love.

## CHAPTER 4 NOTES

p. 26        All quoted material on this page drawn from: Robert Lichello, *Enrico Fermi: Father of the Atomic Bomb,* p. 6.

p. 27        "One boy engages an innocent . . ." Robert Lichello, *Enrico Fermi: Father of the Atomic Bomb,* p. 7.

p. 27        "not to be condoned . . ." through " . . . we teachers cannot keep up with them . . ." Robert Lichello, *Enrico Fermi: Father of the Atomic Bomb,* p. 7.

pp. 27–28    "during the first days . . ." Letter from Enrico Fermi to Enrico Persico translated by E. H. Segre. Pisa. 1918.

p. 28        "dragged into it . . ." Laura Fermi, *Atoms in the Family: My Life with Enrico Fermi,* p. 23.

p. 28        "because it was not easy . . ." Emilio Segre, *Enrico Fermi, Physicist,* pp. 15–16.

p. 29        "I am an ass . . ." Laura Fermi, *Atoms in the Family: My Life with Enrico Fermi, pp. 25–26.*

p. 29        "I have decided to study . . ." Letter from Enrico Fermi to Enrico Persico translated by E. H. Segre. Pisa. 1919.

p. 29        "At the physics department . . ." Emilio Segre, *Enrico Fermi, Physicist,* p. 18.

p. 31        All quoted material on this page drawn from: Laura Fermi, *Atoms in the Family: My Life with Enrico Fermi,* pp. 30–31.

p. 33     "If she wanted something . . ." Emilio Segre, *Enrico Fermi, Physicist,* p. 33.

p. 33     "You shall be goalkeeper . . ." Laura Fermi, *Atoms in the Family: My Life with Enrico Fermi,* p. 4.

# 5

## AH, LOVE! AH, LIZARDS—
## 1925–1928

Laura Capon was slender with long, dark, curly hair. Her face was heart-shaped and her eyes could flash with a certain impudence. Throughout her teens and young womanhood, boys and men found her attractive, even beautiful.

She was also very, very smart. She excelled in school and later at university where she studied advanced mathematics. Laura did not flaunt this intelligence, but neither did she hide it. In 1920s Italy this intimidated many of those same boys and men who were drawn to her.

It would not scare off Enrico Fermi. Nor would she be overwhelmed by his brilliance. At first they would clash as equals. Then they would love as equals and marry as equals. Their life together as husband and wife would be lived as equals.

The family into which Laura had been born was Jewish. They were not devout, but she was raised in the Jewish religion. This did not concern Enrico Fermi at all. But the time would come when Benito Mussolini's anti-Semitic measures would force Enrico and Laura to be concerned.

This anti-Semitism, however, was not immediately apparent. Laura Capon's father was certainly not aware of it. He was a career officer in the Italian navy. He was of a military mind and believed that rules should be followed, laws should be obeyed, and people should comport themselves in an orderly manner. He was upset by the strikes and peasant revolts and demonstrations that followed World War I. He was against socialism and afraid of communism and believed that the government should take any means necessary to crush them and restore order.

Signore Capon welcomed the firm hand of Mussolini at the helm of government. "*Il Duce* knows what he is doing," he assured his family. "It is not for us to judge his actions."

Many Italians felt this way. In those early years, *Il Duce* did bring stability to Italy. He crushed the left and brought an end to rioting and the threat of revolution. Even lifelong railroad expert Alberto Fermi, who opposed Mussolini, had to grant that in the much-quoted phrase of the day, he "made the trains run on time." But the price was absolute control which would lead to conquest, war and, finally, harsh persecution of Italy's Jews.

During those first days his effect was felt most strongly in Rome. In other cities, while the Black Shirts might be in control, tradition was slower in giving way to fascism. This was particularly true in Florence, where in 1925 Enrico Fermi was teaching at the university.

Firenze—Florence—is arguably the most beautiful city in Europe. It is a living monument to the Renaissance and there are more paintings and sculptures by masters in the hallowed buildings and museums and churches which line its cobblestoned streets than anywhere else in the world.

It was a strange place for Enrico Fermi to work out his first major atomic theory and to write the paper which would attract the attention of scientists around the world. The way in which he did it was even stranger. It involved his old prankster friend from his student days, Franco Rasetti.

Rasetti, like Fermi, was teaching at the University of Florence. Always at heart more of a biologist than a physicist, Rasetti had combined his interest in wildlife with his fondness for practical jokes. He and Fermi would hike to a nearby marshland and trap gecko lizards. They would then bring them back to the university dining hall and turn them loose to scare the Florentine peasant girls who waited on tables there.

With one exception, the hikes to the marshland were more important to Fermi than the actual trapping of the lizards. They cleared his mind and he was able to think through problems in physics as he always did when he was relaxing. Beside him a chattering Rasetti might be identifying wildflowers, or pointing out subspecies of insects, but Fermi had long been able to listen to his friend without breaking his own train of thought.

Lately he had been thinking about a recent discovery by Austrian physicist Wolfgang Pauli, a one-time assistant of Max Born's. Pauli, while studying the movements of atomic electrons around a nucleus, had observed that there was only one electron to each orbit. Put another way, no two electrons followed the same path.

Fermi appreciated that this suggested an additional order inside each atom that had not been recognized before. The atoms of gases were more accessible than those of solid forms, so he zeroed in on the movements of atoms and their particles in gases. A long time ago he had imagined a "perfect" gas, a gas that was stable as real gases subject to erratic outside forces are not. Such a perfect gas would be the ideal environment in which to study the behavior of atoms and electrons. The energy given off by radiation could be observed and measured. But what laws would govern such a gas? What would be the connection between them and Pauli's discovery?

The young men had reached the marshland. They had brought with them two long rods, each with a silk noose hanging from its tip. Now they baited the tips of these rods with small insects, much as one might bait a fishing pole. They stretched out at the edge of a muddy pond with their poles extended and waited for hungry geckos to take the bait.

"Gecko dead ahead, sir," hissed Rasetti. "Proceeding north easterly direction. Speed point-oh-one-seven knots."

Fermi spied something Rasetti had not yet noticed. A second, smaller lizard was making for the bait on a path parallel to that of the larger one. They made the final jump at the same moment. The larger gecko took the bait; Rasetti jerked his line, tightened the noose and caught him. But the leaping smaller lizard had fallen short and now it scampered away.

Of course! That was it! No two electrons followed the same orbit because no two electrons were in the same quantum state. In other words, they were not the same size, or not the same temperature, or not charged with the same amount of electrical energy—or all three. These were some of the factors that determined an electron's orbit.

Fermi's mind raced. Already he was thinking in terms of that perfect gas he had dubbed *monatomic gas* and how the orbiting electrons of its atoms might behave. Back in Firenze, he immediately began work on a paper entitled "On the Quantization of a Perfect Monatomic Gas."

The conclusions reached in that paper are still followed by metallurgists and engineers today. Among other things, it presented guidelines to determine how well or badly various metals conduct heat, or electricity. Fermi's calculations were—and are—valid. Not only that, but another scientist, Paul Dirac, had independently come up with the same figures. These formulas are still known as the "Fermi-Dirac Statistics," and the particles of gas which follow their laws are called "fermions."

In recognition of this work, in 1926 Enrico Fermi was appointed a full professor at the University of Rome. Professor Orso Mario Corbino, who was now running the university physics department, also hired Franco Rasetti in a lesser teaching position. Laura Capon was a second-year student of general science at the university.

The previous summer, Fermi and Laura had renewed their acquaintance. They had met through mutual friends in Santa Cristina where both were vacationing, Laura with her family. They saw a lot of each other, going on picnics and hikes with others, or sometimes just by themselves.

Laura liked Enrico much better than she had the first time they had met. She came to realize that because he was attracted to her, he was sometimes nervous, which could cause him to say things that sounded conceited and even pompous. Laura understood, but she could never resist teasing him and puncturing his balloon.

On one occasion, with a group of young people, Enrico, who was slightly older, sounded off: "People can be grouped into four classes," he declared. He defined these as below average intelligence, average—"who, of course appear stupid to us because we are a selected group and used to high standards"—intelligent, and—fourth—"those with exceptional intellectual faculties."

"Aha!" Laura teased him. "Then in Class Four there is one person only: Enrico Fermi."

"You are being mean to me, Miss Capon. You know very well that I place many people in Class Four." But he couldn't resist adding that "I couldn't place myself in Class Three. It wouldn't be fair." When Laura continued poking fun at his conceit, he back-

**1927–1928—Fermi and wife-to-be Laura, before the wedding.** (Author's collection)

tracked. "Class Four is not so exclusive." He tried to mollify her. "You also belong in it."

But she had the last word. "If I am in Class Four," she told him, drawing laughter and applause from the other young people in the group, "then there must be a Class Five in which you and you alone belong."

For the rest of that summer, Fermi was kidded about being the one and only genius in Class Five.

Back in Rome, however, he was "Professor Fermi," and Laura was a student not yet advanced enough for his classes. Publication of the Fermi-Dirac Statistics had made him a celebrity on campus. His brilliance was acknowledged by his fellow teachers as well as by the most advanced students. It was not so easy for Laura to tease him any more.

That year she faced a difficult examination marking the end of a two-year physics course she had taken. There were three examiners. Two of the three were Fermi and Rasetti. The word among students who took the examination before her was that the pair was becoming notorious for their willingness to "flunk with no discrimination."

Laura was afraid she would not live up to their standards. It would be too humiliating to have the one she had teased fail her. The possibility was averted when Fermi and Rasetti became so engrossed in a tennis game that they forgot about the examination. Two less exacting examiners replaced them and Laura easily passed.

Because they had friends in common, Enrico and Laura were always running into one another at off-campus gatherings. Drawn to each other, they would find a corner to engage in conversations. Laura confided to Enrico that she was determined to pursue a career in science and was never going to marry because that would interfere with it.

Enrico, on the other hand, thought that a man should have a wife. He shared his picture of the ideal woman he might marry with Laura. She would have to be physically strong, tall and athletic, and preferably blonde because he found blondes the most attractive. Her family should be of peasant stock, which would ensure that her strength would be lasting. All four of her grandparents should be alive so that any children she and Fermi had would stand a better chance of living a long life.

Laura was of average size, slim, not very athletic and dark-haired. Her family had been city folk for many generations. Her grandparents—all four of them—were dead. Obviously a match between Enrico Fermi and Laura Capon was unthinkable.

Still, they enjoyed each other's company and began spending time together on weekends. Fermi and his friend Rasetti had bought similar Peugot roadsters. They would take Laura and whatever young woman Rasetti was trying to impress at the time for long rides through the countryside in one or another of the cars. Years later Laura would remember those rides as being marked by long waits by the roadside while Enrico and Franco stuck their heads deep under the sports car's raised hood and exchanged opinions about carbon in the valves, or heat expanding the carburetor butterfly and making it stick.

On one occasion, when Enrico and Laura were driving alone to visit an aunt of hers who lived near Florence, the fan belt fell apart. Blushing, Fermi removed the belt from his trousers and rigged a replacement. His resourcefulness impressed Laura more than the Fermi-Dirac Statistics ever had.

During the period of this courtship, which both of them refused to admit *was* a courtship, the Fascists had been tightening their grip on Rome and on Italy. Back on January 3, 1925, when Fermi was in Florence, Mussolini had declared his power absolute and had formally installed himself as dictator of Italy. A short while later he had announced the policy to "Italianize" the nation and the speaking of German and Slavic languages had been forbidden by law. "Pure" Italians had replaced some functionaries in high government positions. Even so, between 1925 and 1928, anti-Semitism was not yet an official part of the party doctrine Mussolini would one day turn into law. Official or not, though, many fascists at all levels of the Party were prejudiced against Jews.

The university was an intellectual oasis from fascism. A center of culture and learning, it did not welcome the absolutes of Fascist doctrine. Still, it was an institution funded by the government and soon there were Fascists among those who ran it, as well as among the students.

They knew who Fermi was, of course. His work had made him the best-known figure on campus. His "romance" with Laura Capon was increasingly gossiped about. Some of the university's anti-Semites were enraged that a young scientist who was sure to bring glory to Fascist Italy should be involved with a Jewish student.

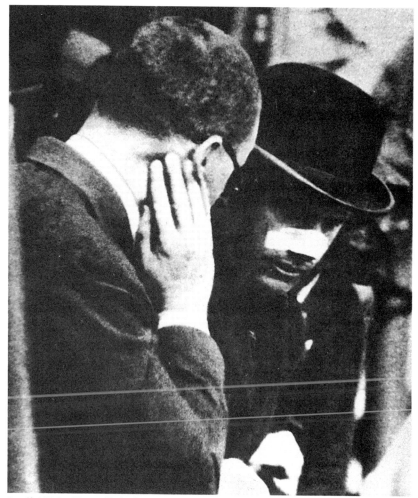

**April 1926—Mussolini after assassination attempt.** (Istituto Italiano di Cultura)

There were mutters of disapproval and occasional remarks made outright. Enrico and Laura ignored them. To them the Fascists were stupid bullies and their insistence on the need to maintain "Italian racial purity" was laughable. Laura teased Fermi that it might interfere with his taking a strapping Nordic type as a wife.

But of course, he had given up this ideal by then. Laura had captivated him. Not just her dark beauty but her quick mind had dazzled him. He had fallen hopelessly in love, and he had good reason to think that she shared his feelings. Enrico asked Laura to marry.

Fascism aside, the fact that they were of different religions—Enrico, Catholic; Laura, Jewish—might have been an obstacle. It was not usual in Italy at that time for people to marry outside of their faith. Besides, Enrico considered himself an "agnostic," and said that he was "not quite sure that God exists." Laura, although not especially pious, did not share that doubt and was somewhat shocked by it. But she faced her feelings. Somehow she had fallen as deeply in love with Enrico as he had with her. She would agree to marry him.

The wedding took place on July 19, 1928. There had been too many barriers to ask a priest, or a rabbi, to perform the ceremony, so it was conducted by a government official in City Hall. The best man was Professor Corbino. Enrico arrived late because the sleeves of his wedding shirt had been 3 inches too long and he had been delayed at the sewing machine shortening them.

Following the wedding, of course, was the honeymoon. It gave everybody pause. Enrico escorted his bride aboard a newfangled two-engine seaplane and they flew to Genoa to begin a new life.

## CHAPTER 5 NOTES

p. 37    "*Il Duce* knows what he . . ." Laura Fermi, *Atoms in the Family: My Life with Enrico Fermi*, pp. 5–6.

p. 38    "Gecko dead ahead . . ." Robert Lichello, *Enrico Fermi: Father of the Atomic Bomb*, p. 9.

p. 39    "Fermi-Dirac Statistics . . ." and "fermions . . ." *Encyclopaedia Americana* (International Edition), Volume 11, 1991, p. 112.

pp. 39–40    "People can be grouped into . . ." through " . . . you alone belong . . ." Laura Fermi, *Atoms in the Family: My Life with Enrico Fermi*, pp. 10–11.

p. 41    "flunk with no discrimination . . ." Laura Fermi, *Atoms in the Family: My Life with Enrico Fermi*, p. 45.

p. 42    "Italianize," *Encyclopaedia Britannica*, Volume 12, 1970, p. 780.

p. 42     "Pure," *Encyclopaedia Britannica,* Volume 12, 1970, p. 783.

p. 43     "Italian racial purity," *Encyclopaedia Britannica,* Volume 12, 1970, p. 780.

p. 44     "agnostic," and "not quite sure . . ." Laura Fermi, *Atoms in the Family: My Life with Enrico Fermi,* p. 108.

# 6

## ATOMS AND GOLF BALLS— 1929–1936

*I swear to be faithful to the King and his royal successors and to the Fascist regime; to observe loyally the statute and the other laws of the state; to exercise the teaching function and to fulfill all academic duties with the purpose of forming citizens active, bold and devoted to the Fatherland and to the Fascist regime. I swear that I do not belong and will never belong to any association or party the activities of which are not in harmony with the duties of my office.*

This Oath of Allegiance to Fascism, along with the demand that all university professors in Italy sign it, was issued on August 28, 1931. It was officially called the "Royal Decree of the Italian Government, Number 1227."

Soon after the decree became public, a group of professors at Harvard University lodged a protest against it with the Institute of Intellectual Cooperation of the League of Nations. Pointing out that it was "applicable to university professors alone," they complained that "this oath involves an intellectual coercion which is incompatible with the ideals of scholars." In France, England, Spain, Switzerland and even Nazi Germany, university professors formed committees to join the protest.

Only 1,225 Italian professors were affected by the loyalty oath. One of these—Professor Vittorio Orlando, the former prime minister of Italy who had headed the peace delegation at the end of World War I—resigned his university position rather than sign the pledge. Eleven others, including Professor Voltera who held the chair of physical mathematics at the University of Rome where Fermi was also a full professor, refused to swear an oath of allegiance to Fascism. Professor Enrico Fermi was not one of the eleven.

During the years following his marriage, Fermi paid little attention to politics. Between his ongoing research into the behavior of atoms and their particles and his active role in building up the University of Rome's School of Theoretical Physics, he had little time left over. Besides, he was a man in love, a man who had just taken on the responsibility of supporting a wife and—soon, no doubt—a family.

In 1929 Enrico Fermi was named to the Royal Academy of Italy, a great honor for one who was only 27 years old. For the inauguration ceremony, he had to wear an elaborate uniform which cost him 7,000 *lire* ($350) to buy. This was more than three times his monthly salary at the university.

As it was, he and Laura were just getting by on that salary. They needed more money. Enrico decided to write a physics textbook for *licei*—Italian high schools— to earn some. It was his idea to "dictate it" to Laura who could "copy it in [her] spare time and help [him] draw sketches for the illustrations."

**Mid-1930s—Members of the Royal Academy of Italy (Academy was presented as a Fascist showcase); Fermi at far right.** (Author's collection)

The young woman he had married, however, was no passive helpmate.

"It is evident . . .," slowly, Enrico would dictate, " . . . that in a non-uniformly accelerated motion the ratio of the speed to the time is not constant."

"It is not 'evident,'" Laura would object.

"It is to anybody with a thinking mind."

"Not to me."

"Because you would refuse to use your brains," Fermi would explode.

But in the end she would prevail. The textbook, after all, was meant for high school students, not for advanced physicists. Laura Fermi provided the simple words to clarify Enrico's writing.

The textbook was not finished when Fermi received an invitation to lecture on quantum theory at the University of Michigan. Laura went with him for a stay which lasted two months. Both of them were favorably impressed by this first visit to the United States.

Back in Italy, they worked to finish the book. By the time it was published, Laura was pregnant with their first child. It was born on January 31, 1931, and they named the infant girl Nella.

Around the time of Nella's birth, Fermi had been working on some very abstract theories in which both physics and mathematics played a part. These theories had to do with the rays given off by nuclei during radiation. He wrote a paper, "Tentative Theory of Beta Rays," which was turned down by the scientific journal to which it was submitted. Today this paper is considered one of Fermi's major accomplishments, but at the time he was depressed that it had been rejected. He decided it would be more rewarding to do concrete experiments than to concoct any more pure theories.

Throughout the early 1930s, Irène Curie and her husband Frédéric Joliot had been working to generate radioactivity artificially. Thirty years earlier her mother, Marie Curie, had observed the "natural" giving off of rays by uranium. These rays were similar to the X rays previously discovered which also gave off what Marie Curie called "radioactivity." Now, Irène Curie and Joliot experimented with bombarding aluminum and other light metals with alpha particles. Alpha particles are the positively charged nuclei of helium atoms. The experiments were successful. They released small particles called positrons. Curie and Joliot had "created" radioactivity.

Impressed by the discovery, Fermi's mind jumped to the possibility of producing artificial radioactivity with neutrons which have no electrical charge. Because they are neither attracted by electrons, nor pushed away by nuclei, neutrons have more energy than alpha particles and they travel faster. Therefore, reasoned Fermi, when they did strike a nucleus, the impact would be much greater. And the greater the impact, the greater the release of radioactivity.

Fermi's first problem was to make sure he had a way to judge the success, or failure, of his experiments. He had to be able to measure the radiation. Only recently, a device had been invented which could do that. It was called a Geiger counter. But there was no such gadget at the University of Rome. It would be much too expensive to buy one, so Fermi had to build his own.

Next he would have to find a source for the ammunition he would need—the neutrons. The best such source, of course, is radium. The smallest quantity of radium, however, was far too expensive. But then Fermi got lucky.

He met a certain Professor Giulio Trabacchi, director of the physics laboratory of the Bureau of Public Health. In the basement of the building in which the bureau was housed, Trabacchi kept a gram of pure radium. He also had a device to extract radon from the radium. This radon would be a direct source of neutrons.

Fermi began to bombard various elements with these neutrons. His homemade Geiger counter showed zero results. Stubbornly, he continued.

Two of his assistants on this project were Franco Rasetti and Emilio Segre. Because when Fermi was concentrating he would snap out orders, they took to calling him The Pope. In more relaxed moments he took this good-naturedly and himself labeled Rasetti The Cardinal and Segre The Basilisk.

Finally the day came when Fermi bombarded flourine with neutrons and the needle of his homemade Geiger counter jumped. He ran tests on still heavier elements, that is, those with denser atomic weight, and these too were successful. Bombarding elements with more atoms generated greater nuclear activity. Finally Fermi experimented with the heaviest element available to him—uranium.

The atomic number of uranium is 92, as there are 92 protons in the nucleus of a uranium atom. Should this atomic number change, should it increase, or decrease, then the element—whether oxygen, or uranium, or any other—would no longer be what it was. It would be a different element and would require a different name.

When Fermi bombarded U 92 with neutrons and analyzed the result, he discovered that the radiation resulted in an element with a new atomic number—93. The implication was inescapable. Fermi seemed to have made a new element, one which had never existed.

The reason it had not existed quickly became obvious. Thirteen minutes after it was created and measured, it disintegrated and vanished. Fermi called this new, short-lived element *ekarhenium.*

Now Fermi was a very cautious scientist. He was careful never to make claims he had not checked and rechecked and then rechecked again. The first report of this element 93, ekarhenium, which he made to the journal *Ricerca Scientifica* in May 1934, was strictly an account of what had been done and of the results which had been observed. Fermi did not claim to have created a new element. But he did report observing a result which had all the properties of a new element.

A month later Professor Corbino—not only Fermi's friend and best man, but also the one who had hired him to do research at the University of Rome—made a speech to an important government group which included the king. He could not resist bragging about "his boys," the brilliant young physicists he had brought to the University of Rome. He told the gathering that "I can conclude that production of this element has already been definitely ascertained." This was understood to mean that for the first time in history a new element had been *created* by man.

Mussolini was delighted. Newspapers under his control heralded "Fascist victories in the field of culture." They crowed that Fermi's discovery "proves once more how in the Fascist atmosphere Italy has resumed her ancient role of teacher and vanguard in all fields."

The news was flashed around the world. A young Italian scientist named Enrico Fermi had discovered a new element. But in

New York, and then in London, skeptical physicists challenged the claim. Fermi tried to say that "numerous and delicate tests must still be performed before production of element Ninety-three is actually proved." Still, he could not deny that there was some evidence that ekarhenium had been created. He could not deny what he had observed.

The argument raged and his reputation suffered from it. How can one prove the creation of something which disappears after 13 minutes? How can one deal with the challenge that what has been observed is really only another form of the original element—uranium 92—and nothing new at all? How can it be confirmed that tiny protons in motion have been counted correctly when the protons quickly disappear? Who can say for sure that one of those tiny protons has not been counted twice?

Eventually, Fermi began to doubt the evidence himself. Five years later, in 1939, he would admit that his "chemical analysis of the results did not enable him to state definitely whether he had proof that the new elements were heavier than uranium, or whether some unusually complicated disintegration was taking place unrecognized." Subsequently, the work leading to the atomic bomb would seem to indicate that a new element really had been created in that early experiment. But more recent evidence has cast new doubts on the existence of ekarhenium. Many scientists believe the dispute will not be resolved for a long time.

In the mid-1930s, though, the Fascist newspapers insisted that any doubts about Fermi's accomplishment were really an attack on Italy. They claimed he was the victim of anti-Italian prejudice. Fermi tried to ignore the charges and counter-charges. Then, in October 1935, a more momentous event pushed the controversy out of the public mind. Italy invaded Ethiopia.

The invasion followed Mussolini's boast that Italy would dominate the Mediterranean, which he called "*mare nostrum,*" meaning literally "our sea." He did not just mean the waters, but also those lands both in Europe and Africa which bordered them. This was in keeping with his grandiose dream of a Greater Italy in which "all must exist inside the state, nothing outside it."

Ethiopia was then an independent country bordering the Red Sea rather than the Mediterranean, but Mussolini knew it was rich in copper, lead, magnesium and iron deposits. A small nation with no army to speak of, Ethiopia was no match for the quarter-million man, well-trained and mechanized Italian invasion force. Still, Ethiopia fought back more fiercely than *Il Duce* had foreseen. Soon Italy had to commit additional manpower and tanks to the struggle just in order to hold the territory it had already seized.

The war was four months old on February 16, 1936, when Enrico and Laura Fermi's second child was born. They named the boy Giulio after Enrico's late beloved brother.

By this time Fermi was deep into a new project involving ways of increasing the release of artificial radioactivity. He was bombarding cylinders of different kinds of metal—lead, silver etc.—with neutrons. The idea was to measure the amounts of radioactivity released by each one as compared to the others. But Fermi stumbled onto a quite different measurement.

He noticed that the amount of radioactivity released by the same element was different depending on where the cylinder was placed. If set on a wooden table, the radiation was greater than if put on a metal table. If silver was placed on lead, it would release less radioactivity than if it was placed on aluminum, and still less than if placed on wood.

Fermi decided to repeat the experiment by putting the cylinder inside the lightest substance he could lay his hands on—paraffin. When he did this, and then brought in his home-made Geiger counter to measure the results, the device seemed to go berserk. The clicks which measured the radioactivity went off rapidly, like a series of exploding firecrackers. Fermi and his assistants reacted:

"Fantastic!"

"Incredible!"

"Black Magic!"

The result was so spectacular that Fermi veered off on another track. Why should a hollow paraffin "holder" produce such an effect? The answer was that paraffin contains a great deal of hydrogen, and in hydrogen the protons are the same size as the neutrons. When the bombarding neutrons hit the hydrogen protons on their way to the silver nuclei, the repeated impact slowed them down without stopping them. These *slow* neutrons were much more likely to collide with a silver nuclei than those which traveled faster.

Fermi likened it to a golfer's putt. A golf ball hit too hard will travel faster and skim over the edge of the hole. One hit slowly will roll right in. The slow-moving neutron was, in effect, rolling right into the silver nuclei. When that happened, large amounts of energy were being released and that energy could be harnessed.

This realization was like a road sign pointing to the splitting of the atom. But the uranium 93 controversy had taught Fermi a lesson. He was not going to make any premature claims. Throughout the next two years he was going to repeat several variations of his experiments until he was sure of the meaning of the results. Then, and only then, would he publish.

Meanwhile, the Ethiopian War was dragging on. Reluctantly, Enrico was forced to consider it. The League of Nations had condemned the Italian invasion and demanded an end to it. Mussolini refused. And when his troops made slow progress, he authorized his bombers to drop poison gas on civilian populations. There were both official and unofficial protests of this barbarism by countries around the world. On May 5, 1936, the Italian army entered Addis Ababa, the capitol of Ethiopia, and *Il Duce* proclaimed victory.

But Fermi was shocked. He wanted nothing but to be left alone in his laboratory, but some things could not be disregarded. The Ethiopian adventure opened his eyes to the regime that claimed his scientific work as a part of their own glory. He had hoped it would be possible to stay aloof from Fascist politics. But then . . .

On October 23, 1936, the Rome-Berlin Axis Pact was signed between Mussolini and Hitler. The Jews of Italy, who knew how the Jews of Germany were suffering under Hitler's Nazis, had cause to fear for their future. Fermi could not ignore this.

Laura, his wife, was Jewish. Their two children, according to racist doctrine, were also Jewish. If that doctrine prevailed in Italy, then their "mixed" marriage could be declared illegal and annulled.

Enrico Fermi had little interest in Fascist politics, but Fascist politics, it seemed, was about to be very interested in him.

## CHAPTER 6 NOTES

p. 46      "I swear to be faithful . . ." *School and Society Magazine*, Vol. 35, No. 889, pp. 47–48.

pp. 47–48  "dictate it" through "use your brains," Laura Fermi, *Atoms in the Family: My Life with Enrico Fermi,* p. 60.

p. 50      All quoted material on this page drawn from Laura Fermi, *Atoms in the Family: My Life with Enrico Fermi* pp. 91–92.

p. 51      "chemical analysis of the results . . ." *Current Biography*, 1945, p. 180.

p. 51      "*mare nostrum*" through " . . . outside it," *Encyclopaedia Britannica,* Volume 15, 1970, pp. 1100–1101.

p. 52      All quoted material on this page drawn from Laura Fermi, *Atoms in the Family: My Life with Enrico Fermi,* p. 98.

# 7

## THE BRASS RING—
## 1938–1939

In May 1938, Adolph Hitler visited Mussolini in Rome. There was a massive parade of Italian military might in his honor. Hitler, whose Nazi army had just seized Austria, reviewed the troops from Mussolini's balcony. A few days later, on May 14, Mussolini declared in a speech that "Fascists will fight together" in the event of war between Germany and any other countries. Not long after Hitler left Rome, Mussolini announced his agreement with Nazi "racist" doctrines and launched a campaign against Italian Jews.

By this time, Enrico Fermi had published the results of his experiments with slow neutrons. They had met with a mixed reaction from scientists around the world. Some heralded a breakthrough in atomic research. Others, remembering the ekarhenium dispute, were slow to accept new discoveries from Fermi.

Fermi had worked hard to prevent another controversy. For almost four years he and his assistants had repeated the first experiments with variations. Since using paraffin had indicated that hydrogen molecules made the difference, he had experimented with elements under water because water also had a large proportion of hydrogen. These further tests had been conducted in a goldfish pond in Professor Corbino's backyard garden. Following them, the research continued with other changes and modifications.

Professor Corbino was practical in a way that Fermi was not. "Don't you understand that your discovery may have industrial applications?" he pointed out. "You mustn't write anything more about this until you take a patent."

Enrico took his advice. He again delayed publishing his results while patents were taken out for a "process to produce artificial radioactivity through slow neutron bombardment." Eight months after the Italian patents were granted, an application was filed in his name at the United States Patent Office. Only then did Fermi publish.

Despite his caution, acceptance of his findings was slow in coming. While Enrico fretted about this, Laura was more concerned with the way fascism was affecting their daughter Nella. One night Laura heard Nella saying her prayers and one of them was not to God, but to Mussolini. When Laura tried to tell her that Mussolini was only a man, and that one did not pray to a person, Nella assured her that "the teacher would not make me say a prayer to Mussolini if he could not hear it."

Nella went to a public school, which meant a Fascist school. Like all the children in the school, she was a member of the Fascist youth organization in charge of the physical education programs. Like them she wore a Fascist athletic uniform. That she was half-Jewish had either gone unnoticed, or was being ignored.

Mussolini's anti-Semitic campaign was not really taking hold in Rome, or throughout Italy. Only one out of a thousand Italians was Jewish. There were large areas of the country where people had never even seen a Jew. Besides, such prejudice was not in keeping with the Italian character or tradition.

Two things changed this, and both alarmed Laura and demanded Enrico's attention. The first was Mussolini issuing the Manifesto della Razza on July 14, 1938. The second was the passage of specific anti-Semitic laws in early September.

The Manifesto della Razza proclaimed that "Jews do not belong to the Italian race." It said that Italians were members of the "Aryan" race, which was defined as a "pure Italian race." It added that Jews "could never be assimilated in Italy because they are constituted of non-European racial elements."

The September Laws incorporated "basic anti-Jewish legislation . . . patterned on the Nürnberg Laws." The Nürnberg Laws were the Nazi legal foundation which deprived Jews of their property and rights of citizenship and eventually were used to send them to the extermination camps. To enforce the September Laws, the Fascists set up an "Institute for the Defense of the Race."

With these laws and the infamous Manifesto, the Fascists succeeded in making anti-Semitism a proof of patriotism for many Italians. The result was an increase in both official and unofficial persecution of Jews.

Enrico Fermi and his family encountered no such persecution. He was not himself Jewish. And although marriages between "Aryans" and Jews was forbidden, and past mixed marriages were nullified, his wife and children were not hassled. He was quietly made aware that his wife's Jewishness would be overlooked. Fermi was a prominent scientist, a credit to Fascist Italy, and it was not in *Il Duce*'s interest to disturb that image.

This was not good enough for Enrico, nor for Laura. They discussed their situation and quickly decided that they and their children must leave their native land. It wasn't just that the government attitude might change if he didn't make any more great discoveries. It was also that when faced directly with Fascist anti-Semitism they realized that by staying they would be condoning it. Fermi was world-famous, and if he allowed *Il Duce* to claim him as a Fascist scientist, a Fascist genius, then by implication he would be approving Fascist policies. It came down to a simple matter of conscience.

Since they had already been to the United States and liked it, Fermi applied for employment to the physics departments of five American universities. Afraid that the Fascists might be watching him, aware that mail going out of the country could be opened by censors, he and Maria took a vacation in the Alps and mailed each of these applications from a different village miles apart.

The replies came quickly. All five universities offered him employment. Fermi decided to accept an offer from Columbia University in New York.

Soon after accepting Columbia's offer, Enrico was confidentially notified that he had been nominated for the Nobel Prize. It was a vindication of his latest work, and whether he won the prize or not, it would go a long way towards ending any doubts about the value of his experiments with neutrons. The merry-go-round of controversy was over and the brass ring was within his reach.

Fermi kept the news of the nomination to himself. It further complicated an already complex financial dilemma. If they emigrated permanently from Italy to the United States, he and his wife would be allowed to take only the equivalent in *lire* of $50 apiece. If he won the Nobel Prize, however, under Italian law he would be required to convert it into *lire* and bring it back to Italy.

The Fermis decided that if he won the prize, they would go to Stockholm to collect it. They would take the children and not return. They would leave for New York from Stockholm. They would use the prize money for travel expenses and to settle themselves in their new life.

But suppose he didn't win the prize? They would still leave Italy. And they would have to find some way to smuggle some money out with them.

On November tenth, Fermi received a call alerting him that the Nobel Prize winners were about to be announced. He was asked to stand by for another call. As he and Laura waited, they listened to the radio.

The news was not good. A new set of anti-Semitic laws was going into effect. Jewish lawyers and doctors could no longer have Christian clients or patients. Jewish teachers were no longer to be employed. Jews could not hire Aryans as servants. Jews no longer had citizenship rights, and their passports were declared void. Jewish children would no longer be allowed to attend public schools.

In theory, this applied to the Fermi children. In theory, Laura's passport was no longer valid. Fermi knew that his prestige still protected his family in these matters, but for how long?

The phone rang. It was Stockholm. The Secretary of the Swedish Academy of Sciences quoted the Academy decision to Fermi over the phone. "To Professor Enrico Fermi of Rome for his identification of new radioactive elements produced by neutron bombardment and his discovery, made in connection with this work, of nuclear reactions effected by slow neutrons," he was awarded the Nobel Prize in Physics. Fermi had won!

Beyond that, the academy had approved his earlier conclusions about creating uranium 93, ekarhenium, as well as his more recent work with neutrons. The award wiped clean the past. And it was a great honor—perhaps the greatest.

But there was a major problem. In 1935 the Nobel Prize had been awarded to an outspoken anti-Nazi German, the author and pacifist Carl von Ossietzky, who at that time was being held in a Nazi prison as "an enemy of the state." Furious, Hitler had barred all Germans from any future participation in the Nobel awards. Now, for an Italian to accept the prize would be an insult to Mussolini's Axis partner.

When he learned that Fermi had won the Nobel Prize, Mussolini wanted to have his cake and eat it too. Party newspapers and magazines crowed that Fascist doctrine had inspired Fermi, and that a Fascist government had helped further his work. At the same time, they denounced the Nobel Prize as a symbol of the decadence of democracies and urged him not to accept it. They went even further. They urged the government to forbid Fermi to accept it.

Mussolini, with one eye on world opinion, would not go that far. Fascist pressure was brought to bear on Enrico privately, but *Il Duce* did not make a public issue of it. Even so, enforcement of the new anti-Semitic laws did throw one monkey wrench into the Fermis' plans.

Laura, like all Jews, was required to turn in her passport. Fermi pulled some strings with a high government official who had always been friendly to him. Two days later Laura's passport was returned with no reference to her religion, no notation as to "race" on it.

On December 6, 1938, the Fermi family left Rome for what would be the last time. Only a few close friends knew that they had no intention of returning. Shortly after they reached Stockholm, Enrico went to the American consulate to do the paperwork necessary to settle permanently to the United States.

There was a problem. The U.S. immigration quota for Italians set strict standards. Little Nella had a vision problem which didn't measure up to them. The sight of her left eye was clouded, and she could see clearly only from her right eye. At the consulate Fermi was told the defect would have to be corrected by an operation before she could be allowed into the United States.

Enrico assured them that he would be able to pay for the necessary surgery after they reached New York. Since he was about to be awarded the Nobel Prize, the consular officials accepted this. The vision requirement for Nella was waived.

But there was no authority to waive the "intelligence test" which Enrico had to take. The immigration officials had to be sure that he was smart enough not to be a burden on the taxpayers. The Nobel Prize was not enough; he had to take the test.

The examiner for immigrants' visas asked him questions like those she asked everybody else: "How much is fifteen plus twenty-seven?"

"Forty-two," Fermi replied with an absolutely straight face.

"How much is twenty-nine divided by two?"

"Fourteen-point-five."

Enrico Fermi passed the test. He would be allowed to immigrate to the United States. Now, with that off his mind, he could accept his Nobel Prize.

It was one of only two prizes awarded by the Nobel Committee in 1938. The other went to Pearl Buck, the American author, in recognition of her novel *The Good Earth* and other works. The two winners sat side-by-side in throne-like chairs with carved lion's heads while one speech followed another from the podium of the Stockholm Concert Hall.

Enrico did not hear much of these speeches. He was too busy concentrating on trying not to breathe, on contracting his chest and holding in his stomach. His over-starched evening shirt was both too tight and too long for his torso. It was in danger of bulging out of the waistband of his formal trousers with any abrupt movement.

Finally the moment came when Fermi was to accept the award—a medal, a diploma and an envelope with a check—from King Gustavus V of Sweden. The king shook his hand, and then Fermi, who had been briefed not to turn away from the royal presence, followed protocol and took four backward steps to his ceremonial chair. Years later he would shake his head in awe at having managed this while at the same time preventing his straining over-starched shirt from flying free of his pants.

Reaction to the newsreel of the event when it was shown in Italy was anything but calm. Fascist journals like *Lavoro Fascista* attacked Fermi for wearing formal "bourgeois" tails similar to those worn by King Gustavus. This was not proper for such a ceremony, they protested. Mussolini would have worn his Fascist uniform, and so Fermi should have worn Fascist regalia regardless of how the king dressed.

**1939—Fermi and Pearl Buck at Nobel Prize ceremony.** (Author's collection)

Worse, Fermi had not given the straight arm Fascist salute when he accepted the award. This was an insult to Fascist Italy. Instead, he had shaken hands with the king. Shaken hands! The Fascist press condemned this as "un-Roman and unmanly." Finally, in keeping with Fascist anti-Semitic doctrine, Fermi's insults to fascism were blamed on the unpatriotic influence of his Jewish wife.

The attack only proved to the Fermis that they were right to leave Italy permanently. Six-and-a-half years later, after the dropping of the atom bomb ended World War II, Prime Minister Winston Churchill would tell the British House of Commons that victory was a result of the Axis powers "exiling, among others, Nobel Prize winner Enrico Fermi." By driving Fermi out, fascist Italy had lost—and the United States had gained—a scientist who would play a key role in the war that was even then on the horizon.

All four Fermis—Enrico, Laura, their daughter Nella who was almost eight years old, and their son Giulio who was not yet three—had their eyes on a different horizon when the ocean liner *Franconia* sailed into New York Harbor on January 2, 1939. They were lined up at the rail, squinting through the early morning fog for their first glimpse of the famous Statue of Liberty. When they spied it, Enrico smiled at his family and spoke:

"We have founded the American branch of the Fermi family," he announced.

---

## CHAPTER 7 NOTES

p. 55        "Fascists will fight" and "racist," *Chronicle of the 20th Century,* p. 482.

p. 55        "Don't you understand . . ." through " . . . slow neutron bombardment . . ." *Pageant Magazine.* February, 1956, pp. 35–36.

p. 56        "the teacher would not . . ." Laura Fermi, *Atoms in the Family: My Life with Enrico Fermi,* p. 101.

p. 56        All quoted material on anti-Semitism on this page drawn from *Encyclopaedia Britannica,* Volume 2, 1970, p. 86.

p. 58        "To Professor Enrico Fermi of Rome . . ." Laura Fermi, *Atoms in the Family: My Life with Enrico Fermi,* p. 123.

p. 59        "an enemy of the state . . ." Laura Fermi, *Atoms in the Family: My Life with Enrico Fermi,* p. 134.

p. 60        All quoted material on this page drawn from Laura Fermi, *Atoms in the Family: My Life with Enrico Fermi,* p. 130.

p. 61        "un-Roman and unmanly," Laura Fermi, *Atoms in the Family: My Life with Enrico Fermi,* p. 134.

p. 62    "exiling among others . . ." *Current Biography,* 1945, p. 179.

p. 62    "We have founded . . ." Laura Fermi, *Atoms in the Family: My Life with Enrico Fermi,* p. 139.

# 8
## AMERICA, AMERICA— 1939–1941

A month or so after his arrival in the United States, Enrico Fermi stood looking out the window of his office high up in the physics tower of Columbia University's Pupin Hall. He did not notice when fellow physicist George Uhlenbeck, with whom he shared the office, came in quietly behind him. It was a bleak and overcast February day, but even so Fermi could see all the way downtown to the buildings at the southern tip of Manhattan Island. The rush-hour streets of New York were teeming with cars and taxis, trucks and buses, and swarms of people. Uhlenbeck watched silently as Fermi cupped his hands to form a small, imaginary ball and looked out over the metropolis.

"A little bomb like that." Fermi spoke aloud. "And it would all disappear."

The words were an about-face for Fermi. At Columbia he had been drawn into the debate about a possible *chain reaction* when an atom was split. It was a continuation of the worldwide controversy sparked by the splitting of the atom on May 2, 1932, by two young British scientists, Sir John Cockcroft and E. T. S. Walton. For seven years the experts had been arguing about the possibility of a chain reaction and its consequences. Would it "blow our planet to pieces," as the *London Daily Mirror* had predicted? Or was the *New York Times* right in following up its headline— "ATOM TORN APART YIELDING 60% MORE ENERGY THAN USED"—with the prophecy that atomic energy would be a boon to industry that would benefit all humankind?

Now, in 1939, physicists at Columbia were studying the question as a practical matter. Fermi had shown that slow neutrons have an increased chance of hitting other atomic particles. Many prominent physicists thought that if they could be slowed down enough, they would provoke a chain reaction. Fermi did not disagree, but he had reservations. He explained them to the Fifth Washington Conference on Theoretical Physics a few weeks after he came to the United States.

Fermi's talk raised grave concerns with two of his colleagues, Leo Szilard and Isidor Isaac Rabi. Three years older than Fermi, Szilard was a Hungarian Jewish refugee from Nazism who had worked with Albert Einstein. Rabi, who would win the Nobel Prize in Physics in 1944, had been born in Galicia and raised in a Yiddish-speaking household on the Lower East Side of New York. With Europe on the brink of World War II, both men felt strongly that Fermi should not have been discussing a chain reaction publicly. Nobody knew how far along the Nazis might be in atomic research, but the energy released in a chain reaction might be used in a weapon that would make them rulers of the world.

Szilard urged Rabi to impress the need for secrecy on Fermi. When Fermi was unresponsive, the two scientists called on him a second time together. "I told you what Szilard thought and you said 'Nuts!'" Rabi reminded Fermi. "Szilard wants to know why you said 'Nuts!'" Fermi replied that there was only a remote possibility of achieving a chain reaction. "What do you mean by remote possibility?" Rabi demanded.

"Well, ten percent," Fermi replied.

"Ten percent is not a remote possibility if it means that we may die of it," Rabi pointed out. "If I have pneumonia and the doctor tells me that there is a remote possibility that I might die, and it's ten percent, I get excited about it."

Szilard agreed with Rabi. Fermi did not. Later Szilard summed up their positions. "Fermi thought that the conservative thing was to play down the possibility" of a chain reaction. "I thought the conservative thing was to assume that it would happen and take all the necessary precautions."

But Fermi was a scientist. It would not be long before his own train of logic would force him to look upon a chain reaction as more and more feasible. He would no longer be able to dismiss the possibility with "Nuts!"

*Nuts!* was his favorite American slang word. As a recent immigrant, Fermi had problems with the English language, as did his family. He was quick at increasing his vocabulary, but his accent clung to him. Often his grasp of words was greater than his ability to make himself understood. He sought help from a graduate student, Herbert Anderson. According to Laura Fermi, the young Anderson "learned physics from Fermi and taught him Americana."

Their first six months in the country, the Fermis lived in New York City, initially at the King's Crown Hotel near Columbia, and later in a furnished apartment rented from a Mrs. Smith whom Laura Fermi called "Mrs. Zmeeth." Her pronunciation was a family joke, but Enrico did not do much better. Finally he consulted Anderson about the language problem.

"He says we should hire our neighbors' children," Enrico reported back to Laura, "and pay them a penny for each of our English mistakes they correct."

This was done with good results although Enrico never completely overcame his accent. Nor—"Nuts!" aside—was he quite sure about some other American slang words. Should eight-year-old Nella and three-year-old Giulio be repeating the words they picked up in the playground—words like "lousy" and "stinky" and "jerk?"

The American lifestyle was a problem too. It was very different from that of Italy. This was brought home to Enrico and Laura one day when Giulio was told by his mother to wash his hands for dinner.

"You can't make me!" the little boy replied. "This is a free country."

By then the family had been living in the free country for awhile. Six months after their arrival Enrico had bought a house in Leonia, New Jersey. Other Columbia scientists also had homes there.

Events had moved quickly during those first months. On March 14, 1939, Hitler's armies had occupied Czechoslovakia. This was a violation of the agreement the German dictator had made with British Prime Minister Chamberlain at Munich only five months earlier. War could break out at any moment.

The occupation followed a key experiment at the Kaiser Wilhelm Institute for Chemistry in Berlin. Here, at the end of 1938, two chemists—Otto Hahn and Fritz Strassman—and a physicist—Lise Meitner, an Austrian Jew who would soon be forced to flee Germany when Hitler annexed Austria—had been working on a chemical analysis of ekarhenium—the controversial "element 93" released in Fermi's bombardment of uranium 92 with neutrons. This analysis showed that the bombarded uranium atoms split into two almost equal parts. Before this atoms had been known to break up and give off protons, but now it was seen that when uranium atoms split equally, an enormous amount of nuclear energy was released. This process was—and is—called *fission*.

It was a breakthrough for the science of physics—a gift from the science of chemistry. It was also a cause of great concern among the scientists, particularly the refugee physicists at Columbia. It told them that Nazi Germany was much further along the path to atomic weaponry than had been thought.

Particularly alarmed, Leo Szilard acted on the news which had been smuggled out of Germany. He performed an experiment with two grams of radium which indicated—in his words—"that the large-scale liberation of atomic energy was just around the corner." It raised both his expectations and his anxiety. "Estimate chances for [chain] reaction now above fifty percent," Szilard concluded.

Two days after the occupation of Czechoslovakia, Szilard, Fermi and Eugene Wigner, "one of the leading theoretical physicists of the twentieth century," called on Professor George B. Pegram, chair of the Physics Department at Columbia. Professor Pegram was quick to grasp the danger of German scientists achieving an atomic chain reaction. It was crucial that the United States develop atomic weapons before the Nazis did. This had to be impressed on the military leaders in the U.S. government.

Enrico Fermi was due in Washington later that day to give a lecture. Because of his prominence as a Nobel Prize winner, Professor Pegram and the others thought he should be the one to voice their urgent concerns to someone in authority. Professor

Pegram wrote a letter to Chief of Naval Operations Admiral S. C. Hooper asking him to meet with Fermi.

The letter was deliberately low-key. Professor Pegram wrote that experiments overseen by Fermi in the physics laboratory at Columbia suggested the possibility "that conditions may be found under which the chemical element uranium may be able to liberate its large excess of atomic energy" to produce "an explosive that would liberate a million times as much energy per pound as any known explosive." He went on to mention Fermi's Nobel Prize and to say that "there is no man more competent in this field of nuclear physics than Professor Fermi."

When Fermi arrived to keep his appointment with Admiral Hooper, he was intercepted by the admiral's aide. The aide asked him to wait and went into the inner office. Fermi then overheard the ethnic slur by which his presence was announced: "There's a wop outside." As Richard Rhodes put it in his book, *The Making Of The Atomic Bomb,* "so much for the authority of the Nobel Prize."

Admiral Hooper had assembled a group of naval officers and officers from the army's Bureau of Ordnance, as well as two civilian scientists attached to the Naval Research Laboratory. They heard Fermi out, and then responded. The army could not see how radiation could be fired from artillery without melting the barrels of the guns. The navy expressed doubts about any sort of container holding such volatile stuff. The civilian scientists, not quite following the intricacies, thought the possibility of a chain reaction too small. Admiral Hooper was polite and asked Fermi to keep the navy informed, but it was obvious that he regarded the whole idea of atomic weaponry as science fiction.

One element—uranium—would be basic in turning that science fiction into reality. It was a scarce resource. The "richest uranium mines in Europe" were in Czechoslovakia. In March of 1939, they had fallen into Hitler's hands.

Concern grew among the Columbia scientists, including Fermi. The Germans had achieved fission with uranium and now they had access to vast supplies of the material. In July of 1939, Leo Szilard and Eugene Wigner went to see the most renowned physicist in the world, Albert Einstein. The three men "tried to evaluate the German progress in atomic research since Hahn's discovery of uranium fission." Their most educated guess was ominous. It was decided that a letter should be sent by Einstein himself to no less a government official than President Franklin D. Roosevelt.

"Some recent work by E. Fermi and L. Szilard," Einstein's letter began, "leads me to expect that the element uranium may be turned into a new and important source of energy in the immediate future." Further on in the letter, the implications were spelled out in no uncertain terms. "This new phenomenon would also lead to the construction of bombs, and it is conceivable—though much less certain—that extremely powerful bombs of a new type may thus be constructed. A single bomb of this type, carried by boat and exploded in a port, might very well destroy the whole port together with some of the surrounding territory."

Einstein's letter did not overlook the practical problems: "The United States has only very poor ores of uranium in moderate quantities," he pointed out.

The message to the president concluded with alarming news: "I understand that Germany has actually stopped the sale of uranium from Czechoslovakian mines which she has taken over. That she should have taken such early action might perhaps be understood on the ground that the son of the German Under-Secretary of State, von Weizsacker, is attached to the Kaiser Wilhelm Institute in Berlin where some of the American work on uranium is now being repeated."

As he signed the letter, Einstein forecast that "for the first time in history men will use energy that does not come from the sun."

Shortly after receiving the letter, President Roosevelt appointed an Advisory Committee on Uranium. But its powers were limited and did not at first include providing money needed by Fermi and his associates to continue their experiments. This hampered them, but it did not stop them. They continued to work on the chain reaction problem.

They needed uranium, and they also needed a material through which the neutrons bombarding it might be filtered to slow them down to 1/1,000 of their initial speed. This was the minimum point at which Fermi computed a chain reaction might be started. The best substance for this purpose was *heavy water,* but none was available. The next best was graphite, but pure graphite was also in short supply.

Heavy water, like ordinary water is made up of oxygen and hydrogen. But the hydrogen molecule in heavy water is twice the weight of the hydrogen molecule in ordinary water. This is what makes it so effective as a filter to slow down bombarding neutrons.

It is not a natural substance. It has to be manufactured. When World War II started with the German invasion of Poland in September 1939, there was only one factory in the world which manufactured heavy water. That factory was located in Vemork in southern Norway.

In April 1940 Nazi armies invaded and quickly conquered Norway and Denmark. The Nobel Prize-winning Danish physicist Niels Bohr, then in his fifties, immediately joined the Danish anti-Nazi resistance movement. He worked actively with them until 1943 when "under threat of immediate arrest, he escaped to Sweden with his family by fishing boat, and eventually went to the U.S."

But in 1940, when the Vemork heavy water plant across the Skagerrak Straits from Denmark fell into German hands, Bohr recognized its importance. At his suggestion the plant was placed under surveillance by the Norwegian underground. Soon Bohr was informed that at the German-held plant "a sinister order was issued to increase its production to 3,000 pounds of heavy water per year!"

When this information was passed on to Fermi and his colleagues, they were frantic. They became obsessed with the fear that the Nazis might develop an atomic bomb before the democracies.

But their fears ran counter to the isolationist feelings of the American people. Although the Axis powers and their sympathizers occupied all of western Europe while England stood alone and vulnerable to an invasion, many Americans, recalling World War I, did not want to be involved. Franklin Roosevelt, up for reelection in 1940, had to moderate his support for England. He had to deal with a Congress slow to vote any funds for the military, let alone for some ivory tower scheme to build a superbomb.

Fermi testified a few times before the Advisory Committee in a vain quest for more money. They found him slightly comic because, according to his wife Laura, "he still spoke with a thick accent and sprinkled whatever he said with a shower of extra vowels." They provided no more financing, and the original $6,000 was quickly spent scrounging up a small amount of uranium and a pitiful supply of impure graphite.

Still, Fermi and his colleagues kept hammering away at the government, determined not to let them forget the urgency of the situation. After almost two years—three years after the Germans achieved uranium fission—this brought results. Vannevar Bush, director of the Office of Scientific Research and Development, notified them that at last an atomic energy research program would be funded.

The Columbia group was elated. They received the news on December 6, 1941. The next day the Japanese air force bombed Pearl Harbor, and the United States government declared Enrico Fermi an enemy alien.

---

## CHAPTER 8 NOTES

p. 64    "A little bomb like that . . ." Richard Rhodes, *The Making of the Atomic Bomb,* 1986, p. 275.

p. 64    "blow our planet . . ." and "ATOM TORN APART . . ." Joan Solomon, *The Structure of Matter,* pp. 133–135.

p. 65    All quoted material on this page drawn from Richard Rhodes, *The Making of the Atomic Bomb,* pp. 280–281.

p. 66    "learned physics from Fermi . . ." and "He says we should hire . . ." Laura Fermi, *Atoms in the Family: My Life with Enrico Fermi,* p. 150.

p. 66    "Mrs. Zmeeth . . ." Laura Fermi, *Atoms in the Family: My Life with Enrico Fermi,.* p. 142.

p. 66    "lousy" and "stinky" and "jerk . . ." Laura Fermi, *Atoms in the Family: My Life with Enrico Fermi,* p. 151.

p. 67    "that the large-scale liberation . . ." through " . . . of the twentieth century . . ." Richard Rhodes, *The Making of the Atomic Bomb,* pp. 291–292.

p. 68    "That conditions . . ." through "than Professor Fermi . . ." Emilio Segre, *Enrico Fermi, Physicist,* p. 111.

p. 68    "There's a wop . . ." Richard Rhodes, *The Making of the Atomic Bomb,* p. 295.

p. 68    "richest uranium mines . . ." Joan Solomon, *The Structure of Matter,* p. 144.

pp. 68–69 "tried to evaluate the German progress . . ." Laura Fermi, *Atoms in the Family: My Life with Enrico Fermi,* p. 165.

p. 69    "Some recent work by E. Fermi . . ." [Einstein letter quote by] Laura Fermi, *Atoms in the Family: My Life with Enrico Fermi,* pp. 165–166 and Robert Lichello, *Enrico Fermi: Father of the Atomic Bomb,* pp. 17–18.

p. 69    "for the first time in history . . ." Laura Fermi, *Atoms in the Family: My Life with Enrico Fermi,* p. 166.

p. 70    "under threat of immediate arrest . . ." *Encyclopaedia Britannica,* Volume 3, 1970, p. 857.

p. 70    "a sinister order . . ." Joan Solomon, *The Structure of Matter,* p. 145.

p. 71    "he still spoke with . . ." Laura Fermi, *Atoms in the Family: My Life with Enrico Fermi,* p. 166.

# 9

## THE CHAIN REACTION— 1941–1942

Enrico Fermi had been constructing the first atomic pile ever made for about six months when he became an enemy alien. The pile was the next step in creating the chain reaction needed to release the energy of the atom. It was made up of layers of graphite alternating with layers of graphite into which tiny uranium chunks had been wedged. It grew day by day.

How large would the pile be when it was completed? Nobody knew, not even Fermi. It would be finished when it reached the size at which a chain reaction would occur. That size was called the *critical mass.*

Really, the pile was built on theory. Fermi was an explorer entering a strange and frightening land. His goal was to achieve a chain reaction, but neither he nor anyone else knew what the results of a chain reaction might be.

The Nobel Prize-winning physicist was a batter at the plate with his eye on the ball rather than on the bleachers where the ball might land if he hit it. He was concerned with the problems at hand, not the problems that solving them might cause. Chief among these was getting the materials needed for the pile with the limited amount of money available. Fortunately, Fermi had Leo Szilard available to deal with this.

It was Szilard who determined the shape of the pile. He had realized that what he called the "lattice" pattern would be "even more favorable from the point of view of a chain reaction than the system of plane uranium layers which was initially considered."

However, the opinionated Szilard did not get along well with Fermi in the laboratory.

Tensions between them eased when Szilard took on the job of scrounging up the materials for the project. High-grade graphite was needed, but no American companies were turning out such a pure product. ("Graphite" is really another name for the black lead used in ordinary pencils; ultimately, the atomic pile in Chicago used enough graphite to supply each person on earth with a lead pencil.) Szilard badgered the Speer Carbon Company in Pennsylvania until they began to make graphite out of petroleum coke. This resulted in a much higher quality of graphite.

It was estimated that to get started "two hundred and fifty tons of graphite of the utmost purity" would be needed and it was recognized that the project would require "much more later." Thanks to Szilard, the first few tons of graphite received at Columbia began to grow slowly to the required amount. He also tracked down the few grams of metallic uranium available in the United States. It wasn't very pure, and nobody knew how that would affect the process of fission, but Fermi decided to work with it anyway. Szilard had zeroed in on another possible source of uranium in the Belgian Congo, but German submarines were a major obstacle to importing it.

Fermi and his assistants set about stacking the blocks of graphite. The task transformed the Nobel Prize winner into a cross between a bricklayer and a coal miner. He and his assistants went home every night covered with soot from the graphite blocks. This was Fermi's situation in the summer of 1941 when he was appointed Chairman of the Theoretical Aspects Subsection of the Uranium Committee.

The pile was being built in a large laboratory room on the seventh floor of the physics building. The room had been stripped bare of all furnishings and equipment for that purpose. Everyone concerned with the project was sworn to secrecy. "SAM, standing for Substitute Alloy Materials, was the code name for Columbia's nuclear laboratory."

At this time one of Fermi's former assistants from the University of Rome, Emilio Segre, whom Fermi had nicknamed The Basilisk, was doing research at the University of California in Berkeley. Like

Fermi, he had fled the fascism of his homeland to enjoy the freedom of the United States. And like Fermi he was working on atoms and fission. The Segre group discovered that by bombarding beryllium with neutrons a substance they called *plutonium 239* was created. Plutonium 239 is a substitute for uranium. It can be easily manufactured and it is relatively pure.

By now the SAM nuclear pile had reached the ceiling. A much larger room with higher ceilings was needed to contain it. There was no such room available at Columbia.

The scientific mind can be devious. Brooding about the space problem, Fermi wondered what would happen if the pile was compressed. This reminded him that graphite is porous; like an ordinary sponge, it is filled with tiny holes which trap air. Air, Fermi knew from his previous experiments, absorbs neutrons before they can reach the target they are bombarding. This loss of neutrons sucked up by air had long been an irritation to Fermi. Now he wondered: What if the air could be removed from the graphite?

To do this, a vacuum (a space in which there is no air) would have to be created. But how? The answer was on the shelves of canned foods in grocery stores and supermarkets.

"'How is air removed in practice?' Enrico asked himself. 'Whenever air is to be taken away from food, the food is inclosed in a tin can. Why not do the same to the pile? Why not can the pile?'"

Fermi had tinsmiths build a room-size can in sections. It was carefully assembled around the pile to make sure it was airtight. Then vacuum pumps were brought in to suck out the air.

The canning of the pile did cut down the neutron loss. This increased the number of neutrons bombarding the pile, but not enough to bring about a chain reaction. Fermi concluded that the pile was still not large enough. The need for a much larger work space now became urgent.

Two accidents briefly distracted Fermi from this need. The first involved Walter Zinn, a Canadian who taught at New York's City College when he wasn't working with Fermi on the atomic project. As he was opening a can of highly flammable thorium in the Columbia lab one day, it exploded in Zinn's hands. He was wearing goggles and rubber gloves, but the gloves caught fire. His hands were seriously burned. The goggles protected his eyes, but not the rest of his face which was left scarred.

Results of the second accident were not appreciated until long after it occurred. It involved two grams of radium which Leo Szilard had obtained from the U.S. government. Fermi and his protégé Herbert Anderson ground up a tiny amount of this radium with beryllium to make a powder they thought might be more practical to work with than beryllium alone.

The mixture was not quite dry and so they put it on a heating slab in a sealed room. When radium is heated it poisons the air. Fermi and Anderson stayed outside the room, but every so often they peeked in to check on the drying process.

One such peek revealed that the mixture was burning and the room was filled with smoke. Fermi and Anderson raced in, turned off the heating slab and raced out, sealing off the room behind them. Since they knew that radium was highly radioactive, they immediately checked themselves with Geiger counters. There were no clicks to indicate that they had been exposed to anything beyond safe amounts of radioactivity.

Some five years later Herbert Anderson began to have trouble breathing. The condition worsened and he became very sick. At first doctors could not identify the cause of his condition. Then they did. Anderson had a rare disease called berylliosis. It had been caused by the beryllium the accident had deposited in his lungs. It was a long time before he overcame this illness.

In those early days, atomic researchers routinely exposed themselves to such risks. It was known that radioactivity was dangerous, of course, but the extent of the danger was not appreciated. Fermi himself could not have said just how much he exposed himself over the course of a lifetime of working with radioactive materials. The incident with burning radium at Columbia was neither the first nor the last such exposure for him.

There was little time to worry about such things now that the United States had entered the war. The atomic research program was speeded up. The government, finally realizing the urgency, put its resources behind it.

The nuts-and-bolts director of the program appointed by the government was Professor Arthur Holly Compton. Like Fermi, he was a Nobel Prize winner in physics. With nuclear projects going on at universities from New York to California and in between, he decided that the first thing to do was to centralize and coordinate them. He picked the University of Chicago as the place to do this.

To Fermi, this meant frequent trips from New York to Chicago. Since his work had top-secret priority, he could not reveal the nature of it to the United States attorney who had to grant permission for each of these trips. "Enemy aliens" like Fermi could not leave their home community without such permission.

John Dunning, a scientist who worked with Fermi at Columbia, remembered just how silly the restrictions could be. One evening he and Fermi were working separately, each on his own research, in the room which housed the Columbia cyclotron. Suddenly Fermi cried out.

"What is it? What's wrong?" Dunning reacted. It occurred to him that Fermi might have discovered something important. "What have you found?"

"It's almost eight o'clock!" Fermi was very upset. He was looking at his watch as if he couldn't believe it.

"Didn't you call Laura to tell her you'd be late for supper?"

"That isn't it. Germans and Italians who aren't naturalized yet are classed as enemy aliens. We can't cross state lines after eight and I live in New Jersey!"

Dunning suggested he call Laura and tell her he was staying overnight at a New York hotel and Fermi seized on the solution. Still, for a man who had good reason to think that the very outcome of the war might depend on his work, the red tape surrounding his vital trips to Chicago was infuriating. On one occasion it was even more so than usual.

Barred from air travel, Fermi was due to leave for Chicago from New York by train. Intricate tests based on his calculations were being run there and one of them had hit a snag. However, his government travel permit, issued in Trenton, had not arrived. Due to leave in the evening, Fermi called Trenton from Columbia that morning. He was told that the permit had been issued and signed, but not sent out. There was no messenger available to get it to him in time. A Columbia secretary raced to Trenton, picked up the permit and raced back to New York, arriving just in time for Fermi's train.

He was furious. "If they want me to travel for them, they'll have to find a way to let me do so freely!" he exploded.

Fermi meant what he said. The government realized this. A permanent permit was arranged which allowed the atomic physicist to travel between New York and Chicago. However, he was still forbidden to fly.

The travel difficulties were eased in April 1942. Work in Chicago had reached the point where Fermi was needed there permanently. He would be in charge of the step-by-step program that would finally lead to a chain reaction.

Because national security was involved and it was important that Fermi's movements attract as little attention as possible, his wife Laura and their children remained behind in Leonia, New Jersey. In Enrico's absence, something happened which made his Leonia neighbors question his sympathies. Some of them, caught up in the war fever of the time, had noted his long absences and worried that the "enemy alien" might be a Fascist spy. These suspicions were heightened by incidents involving five-and-a-half-year-old Giulio Fermi.

When he was a toddler, back in Rome, Giulio had been part of the crowd that witnessed the historic meeting between Benito Mussolini and Adolph Hitler. He had been too young to really remember the meeting, but he had heard it mentioned by his family, and always with the comment that he had been there.

His sister Nella had brought with her from Rome a second-grade reader that had pictures of Mussolini posturing in his black Fascist uniform. When war was declared their parents burned this book, but the heroic pictures had by then become mixed up in Giulio's mind with the meeting he had seen. A small boy in a foreign land, an Italian who spoke with an accent, he seized on Mussolini as a source of pride. He identified with him.

Giulio told the neighbor's children that Mussolini would win the war. The children told their parents and some of them complained indignantly to Laura. Others spread rumors about what Enrico might *really* be up to during his long absences.

Home for a rare weekend, Enrico spoke sternly to Giulio. "I was joking! I didn't mean it," the little boy sobbed.

"Suppose a responsible citizen reports you." Enrico, aware of how strongly Americans felt about the war since the sneak attack on Pearl Harbor, was not exaggerating. "Suppose an FBI man overhears you. What do you think they would do? Wouldn't it be their duty to put you in jail?"

His words to his son were harsh, but Fermi had to assume that the FBI was keeping him and his family under surveillance. The work at the University of Chicago—now code-named The Manhattan Project—was highly secret and everybody connected with it had to be regarded as a possible security risk. Indeed, the more progress that was made, the more dangerous a breach of security could be.

Fermi was a particular concern. His work could give the United States the super weapon which would win the war. Or he could move a decimal point and prevent the development of that weapon. Too much was at stake not to keep a close watch on him and his family.

On November 7, 1942, construction began on a new and more elaborate atomic pile at the University of Chicago. "All calculations and theoretical work had been checked and rechecked" by Fermi and "the danger of an accident or an uncontrollable reaction was practically nil."

Nevertheless, project director Compton worried. He accepted Fermi as a genius, indeed the towering genius in his field. He had to be guided by his judgment. But if the genius Fermi was wrong, the city of Chicago might be reduced to ashes.

Fermi had ruled out that possibility. Looking towards a chain reaction with the new pile, he and his colleagues began referring to it as a *nuclear reactor*. The name stuck. It is still called that today.

The nuclear reactor would be based on Fermi's calculations and built to his specifications. He would be in charge. The assembly would take place under the bleachers at Stagg Field, which, before the war, had been the University of Chicago football stadium.

Stagg was made of concrete and covered with ivy. Under the stands there had been basement lockers and shower rooms and a large racquet court. Now this area would house "Chicago Pile One (CP-1)"—Fermi's nuclear reactor.

The pile was to be shaped like a doorknob about 24 feet in diameter. The graphite blocks would be supported by a wooden cradle to help hold the shape. There would be 40,000 of these blocks, each one 4⅛ inches by 4⅛ inches and 16½ inches long. As at Columbia, the layers of pure graphite would be alternated with layers of blocks into which holes had been bored for uranium chips. The chips were spaced 8¼ inches apart. All of these figures were based on Fermi's calculations—calculations which he had checked and rechecked and rechecked again—and now he would be there to revise and adapt them as necessary during the actual construction.

"Well, Enrico, why don't you lay the cornerstone?" one of the engineers suggested on that fateful November day when construction began. Fermi grinned, hefted a block of graphite and laid it carefully where the chalk marks indicated one of the corners of the nuclear reactor would be. A good-natured cheer went up. They were on their way to a chain reaction.

But the work was as tedious as it was dangerous. The bricks were laid by hand, the layers arranged in a crisscross pattern. Great care had to be taken to align the spaces for the control rods used to keep the layers apart. Otherwise fission might take place prematurely. These rods were "made on the spot: cadmium [carbonate of zinc ore used in electroplating] sheet nailed to a flat wood strip. The strips had to be inserted and removed by hand" in order to measure how close the pile was getting to the fission that would be the first step in the chain reaction. At other times the control rods were locked in place by a simple padlock.

Day-by-day, it was a delicate business. Fermi, according to Herbert Anderson, "spent a good deal of time calculating the most effective location for the various grades of [material] on hand."

By December 1, a little more than three weeks after the construction had begun, it became obvious to Fermi that the pile was approaching the point where a chain reaction could take place. The control rods would have to be removed for this to happen. He observed very closely as each of the rods was taken out. It was late

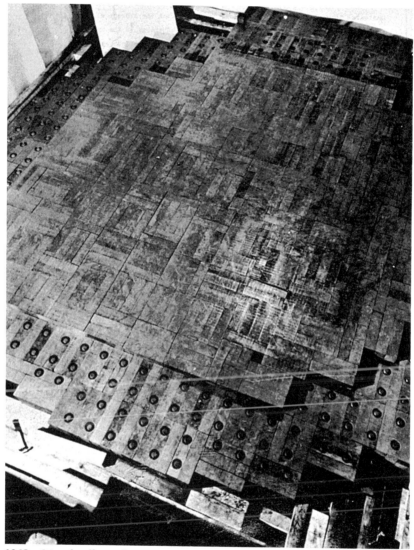

**1942—Atomic pile under construction in the University of Chicago squash court. Chunks of uranium are embedded in the graphite bricks.**
(Author's collection)

in the day and the last rod was left in place. It would be removed very slowly the next morning.

The precautions taken those last two days were so basic they might seem crude. A safety lever the scientists called "ZIP," was set up to disrupt the pile and stop a chain reaction if it got out of hand. It was supposed to operate automatically. But just in case it didn't, a scientist stood by with an axe poised to chop the rope to release it. There was also a "suicide squad, three young physicists installed with jugs of cadmium-sulfate solution near the ceiling." They would douse the pile if it got out of hand.

On the morning of December 2, 1942, under Fermi's supervision, the withdrawal of the last control rod was slowly begun. It was done at 6-inch intervals with long pauses in between. During these pauses measurements of the pile were taken. Fermi made further calculations based on these measurements before ordering further withdrawal of the last control rod.

By 11:30 the rod extended out from the pile about 7 feet. Fermi said "I'm hungry. Let's go to lunch."

When they returned from lunch, Fermi had some further readings taken and worked out some more figures on his slide-rule. "This time," he ordered, "take the control rod out twelve inches." After that was done, Fermi announced that "this is going to do it. Now it will become self-sustaining."

"At first you could hear the sound of the neutron counter." Herbert Anderson described what happened next. "Clickety-clack. Clickety-clack. Then the clicks came more and more rapidly, and after a while they began to merge into a roar . . ."

Fermi's hand shot up and his voice was heard over the noise. "The pile has gone critical!" he announced.

A nuclear chain reaction had begun!

It only lasted four-and-a-half minutes. Then Fermi had the control rod reinserted and the chain reaction slowed down and stopped. The 42 people who were present breathed easy.

So did Professor Compton, who promptly notified Washington. "The Italian navigator has just landed in the new world," he announced. "Everyone landed safe and happy."

## CHAPTER 9 NOTES

p. 73    "lattice" and "even more favorable . . ." Richard Rhodes, *The Making of the Atomic Bomb*, p. 302.

p. 74    "two hundred and fifty tons . . ." Stephane Groueff, *Manhattan Project: The Untold Story of the Making of the Atomic Bomb*, p. 71.

p. 74    "SAM, standing for . . ." Stephane Groueff, *Manhattan Project: The Untold Story of the Making of the Atomic Bomb*, p. 19.

p. 75    "'How is air removed . . ." Laura Fermi, *Atoms in the Family: My Life with Enrico Fermi*, p. 186.

p. 77    All quoted material on this page drawn from Richard Lichello, *Enrico Fermi: Father of the Atomic Bomb*, p. 15.

p. 78    "If they want me to . . ." Laura Fermi, *Atoms in the Family: My Life with Enrico Fermi*, p. 169.

p. 79    "I was joking . . ." through "put you in jail . . ." Laura Fermi, *Atoms in the Family: My Life with Enrico Fermi*, p. 173.

p. 79    "All calculations . . ." Stephane Groueff, *Manhattan Project: The Untold Story of the Making of the Atomic Bomb*, p. 69.

p. 80    "Well, Enrico . . ." Stephane Groueff, *Manhattan Project: The Untold Story of the Making of the Atomic Bomb*, p. 70.

p. 80    "made on the spot . . ." Richard Rhodes, *The Making of the Atomic Bomb*, p. 433.

p. 82    "suicide squad . . ." Richard Rhodes, *The Making of the Atomic Bomb*, p. 433.

p. 82    "I'm hungry . . ." through " . . . self-sustaining . . ." Robert Lichello, *Enrico Fermi: Father of the Atomic Bomb*, p. 21.

p. 82    "At first you could . . ." Richard Rhodes, *The Making of the Atomic Bomb*, pp. 438–440.

p. 82    "The Italian navigator . . ." Richard Rhodes, *The Making of the Atomic Bomb*, p. 442.

# 10

## EXTRAORDINARY MEN—
## 1942–1945

Professor Compton was wrong. Not everyone was happy. Now that Fermi had succeeded, Leo Szilard was having second thoughts. The genie was out of the bottle, and there would be no putting it back. After the chain reaction had been stopped, "I shook hands with Fermi," he would remember later, "and I said I thought this day would go down as a black day in the history of mankind."

A plaque which is still on the outside wall of Stagg Stadium expresses a different sentiment: "ON DECEMBER 2, 1942 MAN ACHIEVED HERE THE FIRST SELF-SUSTAINING CHAIN REACTION AND THEREBY INITIATED THE CONTROLLED RELEASE OF NUCLEAR ENERGY." In the years since it was first produced the arguments between opponents and proponents of nuclear energy have escalated.

At the time though, the war was still on, and so was the pressure to win it. The Nazis had lost the Battle of Stalingrad in which the Russians had killed 90,000 crack German troops, and they were on the run in North Africa as well. But this could mean that they would redouble their efforts to develop an atomic weapon, and that added up to more pressure on Fermi and his colleagues to beat them to it.

They gained time in February of 1943 when a six-man paratrooper commando team of Norwegians was dropped by the British air force onto a frozen lake 30 miles from Vemork. They made their way through heavy snowstorms to the heavy water plant and, although the factory was heavily patrolled, planted explosives under the very noses of the German guards. The detonation that followed spewed half a ton of heavy water into the drains and so wrecked the machinery that the plant was not able to operate normally for over a year. This destruction of heavy water was a serious setback to the Nazi atomic weapons program.

The man who had pressured the British to make this raid was the American Brigadier General Leslie R. Groves. A few months before Fermi's successful chain reaction, the War Department had decided that the Manhattan Project was too important to be left solely to scientists. They picked General (then Colonel) Groves to be in charge of it.

General Groves was not happy about the assignment. He had thought he was in line for an overseas combat command and was disappointed. The last thing he wanted was to play nanny to a bunch of civilian eggheads.

War is the business of generals, and Groves had spent more than 20 years in the military waiting for one to come along to bolster his career. At West Point he had graduated fourth in his class and then had gone on to Army Engineer School, General Staff College and Army War College. At age 46 he stood 5 feet 11 inches and his weight varied between 250 and 300 pounds. His moustache was skimpy but stubborn, and even with his ramrod stiff posture his belly hung over the top of his belt and stuck out below it.

His superiors valued his intelligence and tenacity; his subordinates both respected and feared him. One described Groves as "a tremendous lone wolf." His aide on the Manhattan Project, Lieutenant Colonel Kenneth D. Nichols, thought "he had an ego second to none." Nichols went on to say that Groves "was absolutely ruthless in how he approached a problem to get it done," adding that "I hated his guts, and so did everybody else . . ." He was also fiercely anti-Communist and suspicious of foreigners, and since many of the scientists who worked on developing the atomic bomb were Italian and Jewish immigrants, some with world views far to the left of his own, he would become obsessed with security on the Manhattan Project.

Groves's first security concern was not, however, an immigrant. It was 38-year-old American-born Dr. J. Robert Oppenheimer, the man who would oversee the effort to develop Fermi's chain reaction into a usable military weapon. Groves knew that "the security file on J. Robert Oppenheimer was full of details about his contributions to leftist organizations and also about each acquaintance who was a member of the Communist Party."

Nevertheless, Oppenheimer was both available and had the scientific know-how to take charge of the project. We were fighting a tough war against Nazis, and the only Communist dictatorship around was Russia and it was our ally. Groves reluctantly overruled the FBI and War Department counterintelligence and gave Oppenheimer the job.

Why? What was so special about J. Robert Oppenheimer that the fiercely anti-Communist Groves would champion him? "He's a genius." Groves later explained. "A real genius."

Oppenheimer was not himself a Communist. He was, however, against fascism and troubled about the poor—two concerns that drew him to those on the political left. Tall, stooped, as slender as Groves was burly, often moving in a cloud of tobacco smoke from the pipe between his teeth, he had a sort of befuddled charm that reminded people of the actor Jimmy Stewart.

His soft-spoken manner hid a clarity of vision and a will of steel to match the general's. Where Groves bullied, Oppenheimer got his way by using reason and persuasion. This worked with Groves as it would later with the scientists under Oppenheimer's direction.

Groves thought that security demanded that the scientists be kept separated from one another. He feared "to establish a great university where they discuss their new ideas and try to learn more from each other." Oppenheimer persuaded him of the need to exchange data so that they could build on one another's information. The problems would break down into many different fields—mathematics and theory, chemistry and metallurgy, structural engineering and physics, aerodynamics and meteorology, etc.—and the experts in these fields had to be able to explain their needs and difficulties to each other.

These experts who would turn the chain reaction into a bomb, who would devise a housing for the bomb, who would work out a way to set off the bomb, who would devise the mechanism to move the bomb, who would devise a test and a test-site and who would measure the results—all these experts would have to be

**Albert Einstein and J. Robert Oppenheimer at Princeton.** (New York Public Library; photo by Alfred Eisenstadt)

in the same place. This involved far more people than just the physicists who had come to Chicago. They would have to be brought together and at the same time, for security reasons, they—and their families—would have to be isolated from the outside world. When Groves and Oppenheimer reached agreement on this, the idea for the Los Alamos A-bomb community was born.

To it would come some of the most brilliant scientific minds in the world. First among them would be Enrico Fermi. He was also one of the first—but not the last—to clash with Groves.

The nuclear reactor used in Chicago had been a jerrybuilt affair. It had proved that a chain reaction could occur, but it had not answered the questions about what would happen after it took place. Those questions would have to be dealt with before a bomb could be made.

A much larger nuclear reactor would have to be designed and built. It was a job for industrial engineers. General Groves gave the job to the Du Pont Company. Fermi advised them.

"You're doing all this the wrong way!" was his first criticism. It was relayed to Groves, who took it as proof of just how impractical scientists could be. "You should go out there and build a reactor," Fermi had said. "If it doesn't work, you can find out why it doesn't work, and then you'll find a way to build a second reactor that will work."

But Groves had no patience for failure and second tries. There was a war on. Fermi's slow-moving and cautious logic was too time consuming. Oppenheimer smoothed out their differences as he would other tensions between Groves and the scientists. Given the many strong personalities who would be brought together at Los Alamos, his talent as a referee would come into play many times.

At this point Los Alamos was still in the planning stages. In addition to all of the scientific and engineering structures, there would have to be housing for the families and provision for food and water and electricity and sewage. Schools would be needed, and facilities for recreation. Morale of the families, as well as among the scientists and technicians themselves, would have to be considered. Later, as work proceeded on Los Alamos, more and more requirements were added to the list.

Meanwhile Fermi was traveling between Chicago and a secret location known as Site X in Oak Ridge, Tennessee, where design and construction of apparatus to separate pure from less pure uranium was going on. From Oak Ridge, he would frequently shuttle to Hanford, Washington, where Du Pont was working on the nuclear reactor. Groves and Oppenheimer considered Fermi indispensable, and so Groves laid down security rules for him.

Fermi was not allowed to take a walk by himself in the evening. He could not drive to the Argonne Laboratory 20 miles from his home in Chicago unless he had an escort. And wherever he went, particularly to Oak Ridge, or Hanford, he had to be accompanied by a bodyguard.

The bodyguard was John Baudino, a very tall and bulky Italian-American from Illinois. He was soft-spoken, but his size was intimidating and he took his job very seriously. According to Laura Fermi, during car trips "Enrico sat at the steering wheel, and

Baudino rode by him, a hand on his gun and his face turned back toward the cars behind."

By this time, construction was under way at Los Alamos. On March 15, 1943, Oppenheimer and his aides made arrangements with the hotels in Santa Fe, 34 miles to the south, to house the scientists and technicians and their families who would be arriving over the next month. Most of these experts had been working on different phases of the atomic bomb project, but few of them knew the government's big picture, or how their own work fit into it. With tight security provided by Groves, Oppenheimer arranged lectures to enlighten them. They learned that "the object of the project is to produce a practical military weapon in the form of a bomb in which the energy is released by a fast neutron chain reaction in one or more of the materials known to show nuclear fission."

Among those listening were 30 of the 100 top scientists who would eventually work on the bomb at Los Alamos. The site would also include 6,000 technicians, specialists and soldiers to ensure the security of the area. The support staff would include plumbers and electricians, teachers and librarians, sanitary engineers and groundkeeping personnel. There would even be a barbering staff.

The physicists, though, were the core group. Besides Fermi, whose work already formed much of the basis of the project, many other world-renowned refugee scientists would contribute. Emilio Segre would come to Los Alamos, as would Hans Bethe who would head the Theoretical Physics Division. Bethe, a Jew who fled Germany in 1935, would go on to win the Nobel Prize in Physics in 1967 for "contributions to the theory of nuclear reaction."

One of the older scientists, and one of the more mysterious, was "Mr. Baker." His "eyes were restless and vague; when he talked, only a whisper came out of his mouth." Mr. Baker was actually the Danish underground anti-Nazi fighter Niels Bohr. He was incognito even at Los Alamos with its tight security for two reasons. Although he had recently made his spectacular escape from the Nazi-occupied Denmark, the Germans didn't know this and were still wasting time looking for him there. Also, his wife and two of his sons were still in Denmark and there was a danger of reprisals if the Nazis found out that Bohr was in America working on developing an atom bomb.

**Niels Bohr—Danish anti-Nazi resistance fighter and nuclear scientist.**
(Author's collection)

Less famous was the serious, 30-year-old workaholic theoretician nicknamed "Penny-in-the-slot" because he "only spoke when spoken to," and then gave only one-word answers worth no more than a cent. This dark-haired, round-faced young man with his round spectacles magnifying owlish eyes was a German who had been an underground fighter against the Nazis before fleeing to England where he worked on atomic research for the British government. It would come out later that while at Los Alamos he was passing along crucial information to Russia on the making of the atom bomb. His name was Klaus Fuchs.

Fuchs, during the Fermis' first days at Los Alamos, had volunteered to chauffeur Laura Fermi, who had never learned to drive a car. He was a frequent guest in the Fermis home where he joined in playing charades and other parlor games. In 1950, on trial as a spy, he confessed that "it had been possible for me in one half of my mind to be friends with people, to be close friends, and at the same time to deceive and endanger them."

If Fuchs was destined for infamy, another member of the group, Edward Teller, was bound for the fame and controversy which surround his name right up to the present. Seven years younger than Fermi, the Hungarian-born Teller had green eyes piercing as laser beams, thick, bushy eyebrows that moved with his emotions, firm convictions, and a short fuse when they were challenged. "That young man has imagination," Fermi said of him. "Should he take full advantage of his inventiveness, he will go a long way."

Fermi was right. Teller's contributions to the A-bomb project were major. His dedication sometimes even spilled over into his home life and his time with his small son, Paul. When he taught Paul the alphabet, for instance, he tossed off verses like this:

A stands for atom; it is so small
  No one has ever seen it at all.
B stands for bomb; the bombs are much bigger,
  So, brother, do not be too fast
    on the trigger.

After the war, when Teller devoted himself to developing a hydrogen bomb, many of his colleagues felt that he was indeed being "too fast on the trigger." Up through the end of the Cold War era, Edward Teller was known as "the father of the H- bomb" and as a champion of building bigger and better American nuclear weapons.

All of these men were scientists and independent spirits. From the first they clashed with the military establishment that ran Los Alamos. Oppenheimer was their buffer, but sometimes even he couldn't shield them from Groves.

When the general told his officers that "at great expense we have gathered on this mesa the largest collection of crackpots ever seen," the scientists heard about the speech. Fermi was particularly indignant. "I am an exception," he proclaimed. "I am perfectly normal."

If he was "normal," the life which he and his family and the other scientists at Los Alamos were forced to live was not. Their

**Edward Teller (left) and Fermi.** (Author's Collection)

mail was censored; they were not allowed to have telephones; and nobody—not the wives, or the children—could leave the area without written permission. They called the blocks of tract houses where they had to live "Bathtub Row." The streets were unlit, the roads unpaved and deep in mud. There were frequent plumbing problems and water shortages. Tempers frayed, neighbors argued, family members fought with one another.

Fermi was able to ignore all of this and concentrate on his work. He had been made Associate Director of the project with much more specifically scientific duties than Oppenheimer, and he was the head of F Division. ("F" was for Fermi.) F Division, which included Teller and Herbert Anderson, dealt with problems beyond the scope of the other more narrowly focused groups.

Often this involved Fermi's ability to bridge theory and practice. He was able to make clear to other scientists at Los Alamos how to do this. This led to the solving of one of the most important problems. What kind of trigger would set off the atomic bomb?

Other obstacles had been overcome. The difficulties of how to pack nuclear explosives for rough transport in an aircraft subject to winds and downdrafts had been worked out. A container to keep the elements separated during travel had been designed. So too had a device to time the mixing of them just prior to explosion. The problem of how to make sure the bomb exploded to maximum effect when set off had been solved.

But to be effective, the bomb had to go off while still in the air—before it hit the ground. A trigger had to be devised to do this with split-second accuracy. If the explosion occurred too soon, it would destroy the plane from which it was dropped; too late and its devastating effect would be somewhat limited.

Fermi and Hans Bethe tossed the problem back and forth with the result that a "gun" was devised in the form of a large ball bearing (like one of the metal balls used inside the wheels of roller skates to reduce friction). Holes were drilled in the ball bearing and polonium and beryllium were inserted. Then the holes were plugged with bolts. When the ball bearing "gun" was "fired," it would abruptly mix the elements, releasing a neutron to start the chain reaction in the exploding bomb. The chain reaction would then continue to radiate out over the target area.

Now that there was a trigger, the time had finally come to assemble and test the bomb. People were jittery, tempers short, the tensions between scientists and military personnel greater than ever. A site was chosen for the test at Alamagordo in the southern part of the New Mexican desert. The men who had made the bomb waited impatiently while a steel tower 100 feet high was built and reinforced with concrete. Technicians crossed their fingers as a heavy duty electric winch was used to raise the bomb to the top of the tower. Concrete bunkers with bulletproof glass were set up at what were thought to be safe distances from the target area of the bomb which was called *ground zero*. But nobody could be sure they were really safe.

An hour before sunrise on July 16, 1945, the first atomic bomb in the history of the world was at last exploded. The results were recorded by high-speed cameras. What they revealed was also seen by on-the-spot scientific observer Isidor Rabi:

"Suddenly there was an enormous flash of light, the brightest light I have ever seen . . . It blasted; it pounced; it bored its way right through you. . . . it lasted about two seconds. . . . there was an enormous ball of fire which grew and grew and it rolled as it grew; it

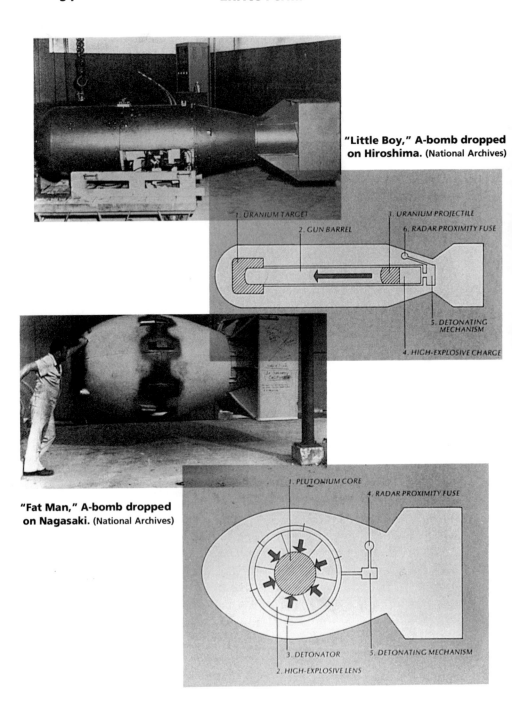

**"Little Boy," A-bomb dropped on Hiroshima.** (National Archives)

1. URANIUM TARGET
2. GUN BARREL
3. URANIUM PROJECTILE
6. RADAR PROXIMITY FUSE
5. DETONATING MECHANISM
4. HIGH-EXPLOSIVE CHARGE

**"Fat Man," A-bomb dropped on Nagasaki.** (National Archives)

1. PLUTONIUM CORE
4. RADAR PROXIMITY FUSE
3. DETONATOR
5. DETONATING MECHANISM
2. HIGH-EXPLOSIVE LENS

went up into the air, in yellow flashes and into scarlet and green. It looked menacing . . . "

Fermi, after measuring off the distances that the shock waves had carried the small strips of paper he had been dropping, quickly calculated that the explosion "corresponds to the blast produced by ten thousand tons of TNT." Although he had no memory of the sound of the explosions, he would later recall his visual impressions: " . . . a very intense flash of light," he would jot down in his notes, "and a sensation of heat on the parts of my body that were exposed . . . the countryside became brighter than in full daylight . . . a conglomeration of flames . . . promptly started rising . . . a huge pillar of smoke with an expanded head like a gigantic mushroom that rose rapidly beyond the clouds, probably of a height to the order of 30,000 feet. . . ."

After the cloud was gone, Fermi and Herbert Anderson climbed aboard two tanks that had been lined with lead as protection against radiation fallout and went to investigate ground zero. The tower, the winch, the steel girders, the concrete supports—all were gone. They had simply disintegrated. Even the asphalt paving had been fused into a shiny green sand that looked like glass but was as hard as diamond. Anderson was able to observe this more closely than Fermi whose tank broke down before he reached the point of impact. Fermi had to walk back to the bunkers.

What were his feelings during that walk? Was he elated at the success based on his work? Did the extent of the forces unleashed make him fearful? We shall never know. Whatever he may have thought before or since, there is no record of how Fermi felt at that critical moment in the history of humanity.

A little more than three weeks later the first atomic bomb was dropped on Hiroshima. A week after that a second nuclear bomb was dropped on Nagasaki. The scope of civilian fatalities caused by the two bombs, most of them women and children, was unprecedented in warfare.

The carnage fulfilled the words spoken by Los Alamos project director Dr. J. Robert Oppenheimer following the test at Alamagordo. In the wake of the fireball, Oppenheimer had quoted the Hindu scripture, the *Bhagavad-Gita:*

**Three nuclear scientists—(left to right) Ernest Lawrence, Fermi and Isidor Rabi.** (Author's collection)

*Now I am become Death, the destroyer of worlds!*

## CHAPTER 10 NOTES

p. 84        "I shook hands with Fermi . . ." Richard Rhodes, *The Making of the Atomic Bomb,* p. 442.

p. 84        "ON DECEMBER 2, 1949 . . ." Laura Fermi, *Atoms in the Family: My Life with Enrico Fermi,* p. ix.

p. 85        "a tremendous lone wolf" through " . . . and so did everybody else . . ." Richard Rhodes, *The Making of the Atomic Bomb,* p. 426.

pp. 85–86    "the security file . . ." Stephane Groueff, *Manhattan Project: The Untold Story of the Making of the Atomic Bomb,* p. 46.

p. 86        "He's a genius . . ." Stephane Groueff, *Manhattan Project: The Untold Story of the Making of the Atomic Bomb,* p. 43.

p. 86        "to establish a great university . . ." Richard Rhodes, *The Making of the Atomic Bomb,* p. 448.

p. 88          "You're doing all this . . ." Stephane Groueff, *Manhattan Project: The Untold Story of the Making of the Atomic Bomb,* p. 133.

pp. 88–89      "Enrico sat at the  . . ." Laura Fermi, *Atoms in the Family: My Life with Enrico Fermi,* p. 213.

p. 89          "the object of the project . . ." Richard Rhodes, *The Making of the Atomic Bomb,* pp. 460–461.

p. 89          "contributions to the theory . . ." *Encyclopaedia Britannica,* Volume 3, 1970, p. 551.

p. 89          "Mr. Baker" and "eyes were restless . . ." Laura Fermi, *Atoms in the Family: My Life with Enrico Fermi,* p. 222.

p. 90          "Penny-in-the-slot" and "only spoke when . . ." Richard Rhodes, *The Making of the Atomic Bomb,* p. 568.

p. 91          "it had been possible . . ." Laura Fermi, *Atoms in the Family: My Life with Enrico Fermi,* p. 210.

p. 91          "That young man has imagination . . ." Laura Fermi, *Atoms in the Family: My Life with Enrico Fermi,* p. 219.

p. 91          "*A stands for atom* . . ." Laura Fermi, *Atoms in the Family: My Life with Enrico Fermi,* p. 222.

p. 91          "too fast . . ." and "father . . ." Richard Rhodes, *The Making of the Atomic Bomb, p. 773*

p. 91          "at great expense . . ." through " . . . perfectly normal . . ." Laura Fermi, *Atoms in the Family: My Life with Enrico Fermi,* p. 226.

p. 92          "Bathtub Row," Laura Fermi, *Atoms in the Family: My Life with Enrico Fermi,* p. 230.

pp. 93–95      "Suddenly there was an . . ." Richard Rhodes, *The Making of the Atomic Bomb,* p. 672.

p. 95          "corresponds to the blast . . ." Stephane Groueff, *Manhattan Project: The Untold Story of the Making of the Atomic Bomb,* pp. 356–357.

p. 96          "Now I am become . . ." Robert Lichello, *Enrico Fermi: Father of the Atomic Bomb,* p. 22.

**August 6, 1945—Mushroom cloud rises from atomic explosion over Hiroshima, Japan.** (National Archives; Dept. of the Army; photo by Sgt. George R. Garon)

# 11
## PROGRESS, OR PORTENT?
## 1944–1946

Enrico Fermi is called "The Father of the Atomic Bomb." It is a distinction that carries a terrible responsibility. "Those scientists who invented the atomic bomb—what did they think would happen if they dropped it?" a fourth-grade Hiroshima student would wonder.

What did happen was that 100,000 people were immediately killed by the atomic bomb dropped on Hiroshima. By the end of 1945, some 40,000 more had died as a direct result of it. Within five years the death toll reached 200,000. Fifty-four percent of the population of Hiroshima perished, and over the more than 45 years since the bomb was dropped, many more have suffered disease and death as a result of the radiation caused by the bomb.

Nor did the devastation stop there. To this day babies are born deformed in Hiroshima as a direct result of their parents'—or even grandparents'—exposure to the radiation unleashed by the bomb.

The effects were similar in Nagasaki. There, the death rate from the bomb was also 54 percent. The initial death toll of 70,000 grew to 140,000 within five agony-filled years. And as in Hiroshima, the effects of the atomic blast continue to this day.

Was use of the bomb justified? The Japanese had bombed Pearl Harbor without warning. They had made war on the United States brutally and relentlessly. Were we not justified in using the most destructive weapon we had against such an enemy?

Fermi thought so. Back in April 1943, afraid that it might take too long to develop an atomic bomb that would work and that the Nazis might beat the Allies to it, he proposed "that radioactive

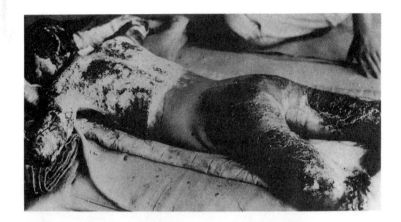

**Atomic bomb victims, Hiroshima.   Top: Backside of burned victim.   Bottom: Injured and bandaged burn victims.** (Japan Society, Inc.; photo by Eiko Hosoe)

fission products bred in a chain-reacting pile might be used to poison the German nation's food and water supply."

The idea was passed along by Oppenheimer to General Groves who discussed it with his military and civilian superiors at the highest level of the government. Finally, Oppenheimer recommended that it be put on hold but not dropped altogether. He wrote Fermi that "we should not attempt a plan unless we can poison food sufficient to kill a half a million men."

How could Fermi support the dropping of the bomb? President Roosevelt had stressed more than once that the nation was engaged in total war, and Enrico Fermi was totally committed to the war effort. He had lived in Italy under Mussolini. He knew that totalitarianism would stop at nothing. He believed it had to be stamped out by any means possible.

Whatever the restrictions placed upon Fermi as an enemy alien by the American government, they couldn't compare to the evils under fascism. To him and his wife Laura, the United States was Utopia compared to fascist Italy. The proudest day of their life was July 11, 1944, when they were sworn in as American citizens in Chicago's District Court of the United States.

Not all of the immigrant scientists agreed that any means was justified to win the war. One who didn't was Leo Szilard. As early as autumn of 1944 Szilard began questioning "the wisdom of testing bombs and using bombs."

These doubts caused General Groves to question Szilard's loyalty. He saw Szilard as "the kind of man that any employer would have fired as a troublemaker." Groves had Szilard placed under surveillance. The counterintelligence report revealed "that Subject . . . has a fondness for delicacies and frequently makes purchases in delicatessen stores." It also noted that he "is inclined to be rather absent minded and eccentric."

It did not mention that Szilard was an intimate of Einstein whose letter to Roosevelt had led to the atomic bomb. In 1945 Einstein gave Szilard a letter to Roosevelt asking that the president give him a hearing. Szilard wanted to be sure the president understood what devastation the bomb might cause before allowing it to be used.

Szilard knew that Eleanor Roosevelt had great influence over her husband, the president. He was able to get an appointment with her for May 8, 1945. The appointment was never kept. President Roosevelt died on April 12.

Harry Truman was the new president. Less than a month after he took office, President Truman appointed a committee headed by Secretary of War Henry L. Stimson to make recommendations on atomic energy. The committee appointed a four-member panel of nuclear scientists to advise it. Fermi and Oppenheimer were two of the members. Among their responsibilities was making "technical recommendations on the military use of atomic weapons against Japan." They were given permission to seek suggestions from other scientists who had worked on the Manhattan Project.

VE Day was proclaimed on May 7 when the Germans surrendered unconditionally. Now the Allies' only enemy was the Japanese. Even though America was winning the war in the Pacific, the general opinion was that the island-hopping campaign being fought would cost a million or more lives before it was over. Work on the bomb at Los Alamos went forward; plans to drop it on Japan were drawn up.

Szilard and six other top scientists recommended to the Fermi-Oppenheimer panel on June 11 that the Japanese be shown a demonstration of the bomb's power instead of dropping it on them. They were overruled. "We can propose no technical demonstration likely to bring an end to the war; we can see no acceptable alternative to direct military use," was the judgment of Fermi and his three colleagues.

In the meantime, Szilard managed to see James Byrnes, then an unofficial advisor to President Harry Truman, but soon to be his secretary of state. The meeting was preceded by Szilard's memorandum pointing out that the A-bomb program was pushing the United States "along a road leading to the destruction of the strong position (it) hitherto occupied in the world." It correctly predicted that "the greatest immediate danger which faces us is the probability that our 'demonstration' of atomic bombs will precipitate a race in the production of these devices between the United States and Russia."

Byrnes was not impressed. He pointed out that two billion dollars had already been spent on the bomb "and Congress would want to know what we had got for the money." He added that dropping the bomb on the Japanese would "make Russia more manageable in Europe."

Secretary of War Stimson was slightly more cautious than Byrnes. He had gone over the recommendations by Fermi and the other scientific panel members very carefully. On July 2 he recommended to President Truman that "a carefully timed warning be given to Japan."

During the next two weeks, work continued at Los Alamos. It was a time of great pressure on Fermi, the man who would reap the glory if the test went well, and possibly the blame if it didn't. The night before the trial, Fermi had tried to get the other scientists to make bets with him on whether or not the explosion would "ignite the atmosphere, and if so, whether it would merely destroy New Mexico, or destroy the world." Later, when they saw him tearing up his bits of paper to measure the aftershock, they thought him "eccentric, or perhaps mad."

Ten days after the Alamagordo test, President Truman issued the Potsdam Declaration. Stimson had overseen the writing, which threatened "the utter devastation of the Japanese homeland." It called for unconditional surrender, but allowed for a "constitutional monarchy under the present dynasty." In other words, the Japanese would be allowed to keep their emperor, a condition which U.S. intelligence sources said was the only obstacle to ending the war. President Truman took the concession out before releasing the proclamation.

The Potsdam Declaration aroused Leo Szilard to further action. He wrote a petition and persuaded 69 of his fellow atomic scientists to sign it. The petition said that if the United States used the bomb, it would "bear the responsibility of opening the door to an era of devastation on an unimaginable scale."

Fermi disagreed. Laura Fermi summed up his attitude: "Enrico did not think that for the physicists to stop their work would have been a sensible solution. Nothing is served by trying to halt the progress of science. Whatever the future holds for mankind, however unpleasant it may be, we must accept it, for knowledge is always better than ignorance."

He believed that if the bomb had not been made, if the research had been destroyed, other scientists would soon have come along to duplicate the experiments and build it. If that had happened, it might have come under the control of a truly evil government.

Besides, the bomb would save lives. Like most Americans Fermi thought there would be a million or more casualties on both sides if the war had to be fought island by island. In many battles the

**Leo Szilard testifies before joint congressional committee.** *(News Front Magazine)*

Japanese had already proved themselves willing to die rather than surrender. (Yet by now there were truce overtures.)

Many United States military leaders, particularly many of whom were battlefield commanders in the Pacific, approved the dropping of the atom bomb as a means of bringing the war to a quick end and avoiding battlefield casualties. A few military historians still hold this view. However, since the declassification of August 1945 diplomatic communications between Japan and the United States, strong doubts have been raised regarding the need to have bombed Hiroshima and Nagasaki.

In the immediate aftermath of the dropping of the bomb, some of our top military leaders questioned its necessity. General (later President) Dwight Eisenhower thought "the Japanese were ready to surrender and it wasn't necessary to hit them with that awful thing." The Supreme Commander of Allied Forces in the Pacific, General Douglas MacArthur, did not think the bomb had been "of any military use against Japan." U.S. Air Force General "Hap" Arnold believed the war could have been won by conventional bombing. Admiral E. J. King thought a naval blockade would have done the job without involving loss of life by land troops. Admiral Leahy thought the bomb was dropped only "because of the vast sums that had been spent on the project."

There was—and is—honest disagreement. As a tactical matter justification for having used the bomb boils down to what was believed about the condition of the Japanese war machine at the time. Had the Japanese fleet been destroyed by August 1945, or was it lying in wait to ambush our navy? Were their planes virtually unable to fly, or did their kamikaze reserves represent a danger equal to the havoc they had wreaked in the past? Is it true that American aircraft were bombing Japanese war plants and munitions dumps at will, or were our bombers taking significant losses on their runs over Japan? Was the Japanese Army really on the run, or was it regrouping to repel an invasion?

Certainly Fermi and his fellow scientists had no way of knowing the military situation. They had made a weapon to win a war and thought it should be used. Emilio Segre, who had worked on the bomb with Fermi, agreed with Fermi's position on this even after Hiroshima. But he did express one doubt: "Whether, after Hiroshima," Segre wrote, "the Nagasaki bomb was dropped too early is another question."

Following Nagasaki, President Truman declared that "we have spent two billion dollars on the greatest scientific gamble in history—and won." So we had. On September 2, the Japanese gave up. What was called an "unconditional surrender," really wasn't. Despite President Truman's having taken the provision out of the Potsdam Declaration, the Japanese were allowed to keep their emperor.

Twelve days earlier, on August 21, 1945, an accident took place in an isolated laboratory at Los Alamos that allowed Fermi to see for himself what atomic energy could do to human flesh. At the time, the accident was kept secret. It would be seven years before the details were made public.

Late in the evening, after normal working hours, two technicians went back to the lab to check on a *critical assembly*. A critical assembly is an atomic pile, or nuclear reactor, which is carefully maintained at just below the point where a chain reaction can start. The pair were gathering information on atomic activity going on at that crucial stage.

Suddenly the pile went out of control. A chain reaction started. One of the technicians, Harry (his full name has not been released), age 26, had his hands on the reactor when the accident occurred.

The chain reaction was quickly brought under control. Both men were rushed to the Los Alamos hospital. Harry's hands were swelling rapidly.

Tests revealed that Harry had "received over two hundred thousand times the average daily dose [of radiation] to which men working with radioactive materials would normally be exposed." There were hundreds of thousands of cases of radiation poisoning in Japan in the wake of the atomic bombs that had been dropped, but this was the first at Los Alamos. Fermi followed it closely and his wife, Laura, went with him to study the pictures taken at intervals of Harry's hands.

They showed "the rapidly increasing deterioration and the painfulness of his condition," according to Laura Fermi. She described "huge blisters, loss of skin, effects of poor circulation in his fingers, finally gangrene."

Harry was in agony. Drugs did not help very much. Twenty-four days after the accident, he died.

The horror of this death, so close to home, impressed Fermi. He was also affected by the pope's condemnation of the atom bomb

**Nagasaki, 1945—Silhouettes of ladder and human figure vaporized by atomic blast are left on a wall.** (U.S. Air Force; photo by Eiichi Matsumoto)

following the Nagasaki raid. And then there came a letter from Rome, from his beloved older sister, Maria Fermi:

" . . . Everyone is most dismayed by its terrible effects," Maria wrote of the atom bomb. "And as time passes their worry increases rather than gets less. I pray to God for you. He alone can pass moral judgement on you."

More and more of the scientists who had worked on the bomb were also sickened by the devastation. In January 1946 at Los Alamos a group of them formed the Association of Men of Science to try to prevent such an atrocity from ever happening again. Fermi did not join the association. It was not just that he and his family had already moved from Los Alamos to Chicago, where he would teach at the university. It was also that he did not agree with their demand that "an international conference be called and a world authority for the control of nuclear energy be established."

Fermi saw the demand as a step towards world government and he did not think that humanity was ready for that. He believed that the horrors of war were not caused by the creation of deadlier weapons, but rather by the drive to use them. He thought that drive was part of human nature and would not change.

**General Leslie Groves pins Congressional Medal of Merit on Enrico Fermi. Other scientists receiving medals are (left to right) Harold Urey, Samuel Allison, Cyril Smith and Robert S. Stone.** (Author's collection)

A short time later the U.S. government awarded Fermi the Congressional Medal of Merit. It was given to him for "exceptionally meritorious services . . . in the making of the most powerful military weapon of all time, the atomic bomb."

Fermi was proud to accept the award. He still believed that nothing should "halt the progress of science." He still thought that "whatever the future holds for mankind, however unpleasant it may be, we must accept it, for knowledge is always better than ignorance."

## CHAPTER 11 NOTES

p. 99      "The Father of the Atomic Bomb . . ."*New York Times,* Nov. 28, 1954, p. 25.

p. 99      "Those scientists who . . ." Richard Rhodes, *The Making of the Atomic Bomb,* p. 734.

pp. 99–101      "that radioactive fission . . ." and "we should not attempt . . ." Richard Rhodes, *The Making of the Atomic Bomb,* p. 510.

p. 101      "the wisdom of . . ." Richard Rhodes, *The Making of the Atomic Bomb,* p. 635.

p. 101      "the kind of man . . . " through "absent minded and eccentric . . ." Richard Rhodes, *The Making of the Atomic Bomb,* p. 506–507.

p. 102      "We can propose no . . . " Emilio Segre, *Enrico Fermi, Physicist,* p. 152.

p. 102      "along a road leading . . ." Richard Rhodes, *The Making of the Atomic Bomb,* p. 637–638.

p. 102      "and Congress would want . . ." through "more manageable in Europe . . ." Charles L., Jr. Mee, *Meeting at Potsdam,* p. 16. (Also, *The Making of the Atomic Bomb,* p. 638.)

p. 103      "a carefully timed warning . . ." Charles L., Jr. Mee, *Meeting at Potsdam,* p. 61. (Also Stephane Groueff, *Manhattan Project,* p. 352.)

p. 103      "ignite the atmosphere . . ." through " . . . perhaps mad . . ." Richard Rhodes, *The Making of the Atomic Bomb,* p. 684.

p. 103     "the utter devastation . . ." and "constitutional mon-archy . . ." Charles L. Mee, Jr., *Meeting at Potsdam*, pp. 16–17 and p. 218.

p. 103     "bear the responsibility . . ." through "always better than ignorance . . ." Pierre De Latil, *Enrico Fermi, The Man and His Theories*. (Translated from the French by Len Ortzen) p. 127.

p. 103     "the Japanese . . ." and "because of the vast sums . . ." Charles L. Mee, Jr., *Meeting at Potsdam*, p. 62 and p. 192. (Also in Richard Rhodes, *The Making of the Atomic Bomb*, p. 688.)

p. 105     "Whether after Hiroshima . . ." Emilio Segre, *Enrico Fermi, Physicist*, p. 155.

pp. 105–106 "we have spent two billion . . ." William Miller, *A New History of the United States*, p. 418.

p. 106     "received over two hundred thousand times the av-erage . . ." Pierre De Latil, *Enrico Fermi, The Man and His Theories*, pp. 127–128.

p. 106     "the rapidly increasing . . ." Laura Fermi, *Atoms in the Family: My Life with Enrico Fermi*, p. 233.

pp. 107–108 " . . . Everyone is most dismayed . . ." Laura Fermi, *Atoms in the Family: My Life with Enrico Fermi*, p. 245.

p. 109     All quoted material on this page drawn from Pierre De Latil, *Enrico Fermi, The Man and His Theories*, p. 128 and 130.

# 12

# THE FINAL SACRIFICE— 1946–1954

The end of the war changed things for Enrico Fermi and his family. On New Years Eve of 1945, they moved back to Chicago. Fermi began teaching at the University of Chicago's government subsidized Institute of Nuclear Studies. He also did research which would later be part of the planning for the development of a giant cyclotron the institute would build.

**1951—An early, 60-inch cyclotron of the University of Chicago type.**
(Brookhaven National Laboratory)

**1952—The Cosmotron, a more advanced cyclotron.**
(Brookhaven National Laboratory)

The cyclotron, a machine to energize atomic particles through magnetic action, was to be part of the government's "Atoms For Peace" program, which was supposed to improve the lives of Americans. People were not yet concerned about the nuclear arms race. We were still the only country that had the atom bomb. General Groves was sure "it would be five to twenty years before even the most powerful of nations could catch up with the U.S. in atomic bomb development." However, it was only three years later, on September 22, 1949, that Russia set off its first atomic bomb.

Fermi was intrigued by the cyclotron project, but he was also caught up in teaching. He started out giving intricate lectures and found that many of his students had trouble following them. This made him realize that if students are to understand advanced physics, then basic science courses must be very well taught.

That autumn, the Nobel Prize–winning Fermi began teaching a course in elementary physics. He approached this as a problem to be solved in much the same way he solved problems in the laboratory. "One must take more time preparing" when teaching the basics, he decided. "I have small puzzles sometimes finding a way to present something so they can understand. This little game I find most interesting. . . . I look into their faces and if I see more bewilderment than one usually finds in the faces of young students at eight-thirty A. M., then I begin all over."

His method now was different than it had been when he taught in Rome. He was older, more tolerant and more patient. If students had trouble learning, he didn't blame them. He thought up different ways to make them learn.

Once or twice a week he lectured to a small group of graduate students. Chinese physicist C. N. Yang was one of these. "Fermi gave extremely lucid lectures," he recalls. "Physics is to be built from the ground up, brick by brick, layer by layer," was Fermi's advice. "I remember his emphasizing that as a young man one should devote most of one's time attacking simple, practical problems rather than deep, fundamental ones."

**1990—The Alternating-Gradient Synchrotron, a modern-day cyclotron.** (Brookhaven National Laboratory)

Fermi liked the quiet life of the university, but he was too prominent to be confined to it. On August 1, 1946, President Truman signed the Atomic Energy Act into law. This created the Atomic Energy Commission (AEC), which had two advisory groups, the Military Liaison Committee to review questions of national defense, and the General Advisory Committee (GAC), which was made up of scientists and engineers to guide the AEC on technical questions. Fermi served on the GAC from January 1, 1947 to August 1, 1950.

This was not a position he sought. He accepted it out of a sense of duty to his adopted country. And in cold, hard cash, it cost him dearly.

In the 1930s, at the urging of his mentor, Professor Corbino, Fermi had joined with those who worked with him on the slow neutron experiments to take out a patent in the United States. The patent covered "the fundamental process for generation of atomic power." This was the basis for the development of the atomic bomb, and now would be the basis for much of what American corporations were proposing to do under the Atoms For Peace program.

While the war was on, and secrecy was crucial, Fermi had not tried to enforce these patents. All his energies were devoted to making sure that it was the United States which benefited from his discovery, and not the Axis powers. But now, with the war over and a future looming in which his work (according to *Pageant Magazine* in 1956) "has kept the A-fuel burning in every atomic furnace that has since been built," Fermi felt that he should share in the profits.

His position as an advisor to the Atomic Energy Commission kept him from doing this. It was the United States government that provided his technology to the companies now using it. To enforce his patents, he would have to sue the government. But because of his AEC advisory job, Fermi was technically a government employee and so, Justice Department lawyers told him, he could not sue the government.

When his term expired in 1950, Fermi did file suit. But by then there was another complication. Bruno Pontecorvo who had played a minor role in the slow neutron experiments in Rome, had generously been included by Fermi as one of the holders of the U.S. patent. He was a party to Fermi's suit. While

**Late 1940s—Fermi at the blackboard.** (Author's collection)

it was pending, Pontecorvo defected to Russia. Without him, the legality of the suit was called into question.

Three years later, in 1953, a settlement was negotiated with the AEC. After lawyers' fees, Fermi was left with about $40,000—small change compared to the millions of dollars which were being made on his discovery. A year later the AEC awarded him another $25,000 in recognition of his contribution to the atomic energy program.

Throughout this time, Fermi had been involved in another controversy that actually had its beginnings in March 1941. Back then, over lunch with Edward Teller, Fermi had wondered idly if an atom bomb might be used as a *trigger* to set off a much more powerful device. This device would be based on *thermonuclear fusion*—the use of extreme heat to melt down hydrogen into helium—and would one day be known as a hydrogen bomb. Teller thought it over and a few days later "I explained to Enrico why a hydrogen bomb could never be made."

Later Teller changed his mind. At Los Alamos he thought development of an H-bomb should be "among its objectives." Fermi and Oppenheimer and most of the other scientists considered this an unwise detour from the race to create an atom bomb. However, Oppenheimer believed that after the war "violent thermonuclear reactions (should) be pursued with vigor and diligence, and promptly."

At first Fermi agreed. Following the success of the Alamagordo test, he and Teller began working on theories of how to use an atom bomb to trigger an H-bomb. When Teller followed Fermi to the University of Chicago in 1946, he continued working on H-bomb problems and from time to time Fermi helped him with them.

In 1949, Teller went back to Los Alamos. His native Hungary was under Russian Communist control. His parents were Jewish and had survived the Nazis, but now they could not be reached. Edward Teller became very concerned about the threat to the free world by the Soviet Union under Stalin. This concern grew in September when the Russians exploded their first atom bomb. His main fear was that they would build a hydrogen bomb before the United States did.

Fermi saw a greater danger, as did Oppenheimer. Fermi's work with Teller and on his own had made him rethink his convictions about science and progress. As a member of the GAC, he joined in the unanimous recommendation to the AEC that a crash program to develop an H-bomb *not* be undertaken. He felt so strongly that he joined with Isidor Rabi in submitting a minority report outlining the reasons for his change of heart:

*The fact that no limits exist to the destructiveness of this weapon makes its very existence and the knowledge of its construction a danger to humanity as a whole. It is necessarily an evil thing considered in any light. For these reasons, we believe it is important for the President of the United States to tell the American public and the world that we think it is wrong on fundamental ethical principles to initiate the development of such a weapon.*

The GAC's recommendation was dismissed. Edward Teller prevailed. He convinced the government that a Russian H-bomb was a real danger. On January 31, 1950, President Truman announced financing for an H-bomb program.

In a way, Teller and Fermi were both right. The first U.S. H-bomb was set off underground and in secret in November 1952. Its force was the equivalent of 7 million tons of TNT. "This explosion caused an island to disappear and created a hole in its place 175 ft. deep and a mile in diameter."

In August 1953, the Russians set off their underground H-bomb. Now the problem for them, as for us, was how to package it for delivery by a plane and how to set it off before impact without destroying the bomber. Secretary of Defense Charles E. Wilson believed the Russians were "three or four years behind where we are" in working this out. He was wrong. They exploded an H-bomb in November 1955, while ours wasn't set off until six months later.

While Teller had been working on the H-bomb program, Fermi had turned his attention to the University of Chicago cyclotron. Begun in 1947, it was completed in 1951. Fermi immediately became involved with Herbert Anderson in running experiments with it.

A cyclotron is a complex device that looks deceptively simple. The one at Chicago had two main parts described by Laura Fermi as "a huge magnet and a metal box . . . so large that it could be used to store 300 bushels of grain." The air was removed from this box by nine vacuum pumps. Then a succession of various elements was put into the box. These were bombarded by protons. Radio frequencies were used to make the protons move faster while the giant magnet, which weighed 2,200 tons and was turned on by an electric current, directed their movements towards the target elements.

**1953—Enrico and Laura Fermi.** (Author's collection)

There was, however, a problem. With the magnet turned on, the cyclotron was radioactive. Nobody could go near it. In order to change the path of the bombarding protons, or to substitute one target element for another, the cyclotron had to be turned off. This wasted a lot of time, and it also created certain dangers.

When the magnet was on, it attracted metal objects *outside* the cyclotron. It pulled the keys from the men's pockets, and hairpins from the women scientists' hair. This was amusing, but then something happened which wasn't.

**1950—Chicago cyclotron.**  (Author's collection)

One day Herbert Anderson noticed a piece of concrete on the floor of the huge room that housed the cyclotron. He went to remove it, not knowing that it was reinforced with steel. As soon as he lifted it, the magnet "grabbed" the fragment. As a result Anderson's hand was badly crushed between the concrete and the housing of the magnet.

Now Fermi saw the magnet as a challenge. He studied it and then came up with an idea for controlling it which would use its own power to do so. The cyclotron would not have to be shut down while tests were altered, or old tests scrapped and new ones begun.

He designed a "trolley" made of lucite. It was a simple platform sitting on four metal wheels. It rode the lower magnet pole inside the vacuum box in the same way as a train rides rails. It could move the target element, replace it or act to redirect the proton bombardment. It was powered by the magnet's field and controlled by an operator using push-buttons in a control room.

**1951—"Fermi's Streetcar," hand-built for Chicago cyclotron.** (Author's collection)

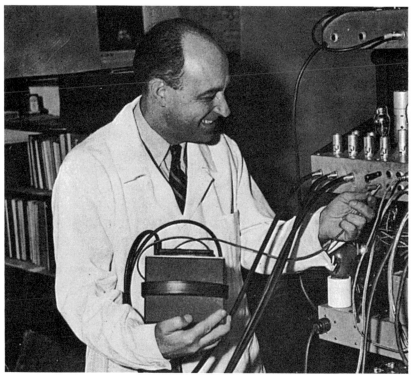

**Eary 1950s—Fermi fiddles with one of his "gadgets" (possibly the "Streetcar").** (Author's collection)

Fermi put together this gadget with his own two hands. It looked like something built from an erector set by a child not quite up to the job. Wires stuck out of it at wild angles. It looked very, very sloppy—but it worked. Today, prettied up and looking very high-tech, *Fermi's Trolley* is still used inside the chambers of cyclotrons.

In 1953, Fermi finished his experiments with the Chicago cyclotron. In the summer he went to Los Alamos to put his notes in order and analyze them. After that he turned his attention to a new subject—cosmic rays. These are submicroscopic electrically charged particles that travel through space at speeds approaching the speed of light. When they enter the earth's atmosphere they collide with other particles and transfer energy to them. These new particles are called *secondary* cosmic rays. They are ultimately converted into heat that may be felt on the surface of the earth. Fermi wrote papers speculating on the origin of *primary* cosmic rays.

During this period Fermi had not been feeling well. He had stomach problems and was losing weight. He devoted himself to his work and tried to ignore the pain.

Not pain, but politics interfered with scientific investigations in late 1953. That was the peak of an era which had begun shortly after the war. The United States and Russia had become locked in a Cold War—so called because while direct confrontation was avoided, there was an ongoing ideological struggle between democracy and communism, along with a nuclear arms race and frequent armed conflicts—some minor, some major—on many fronts. In the United States, the fear of communism abroad led to the fear of communism at home. This danger was exploited by some politicians and journalists and others for personal advantage. Chief among these was Wisconsin Senator Joseph McCarthy whose name would become a byword for the dubious zeal with which suspected communists were pursued in the United States.

McCarthyism clouded the political lines. First it blurred the distinctions between communism and socialism, then between socialism and anti-fascism, then between anti-fascism and liberalism, and finally between liberalism and those who believed in various New Deal reforms of the Democratic Party. So great was the fear of Communist infiltration that teachers, government employees and even factory workers were forced to sign loyalty oaths. Many people lost their jobs and many reputations were ruined.

One victim of McCarthyism was J. Robert Oppenheimer. In November 1953, L. W. Borden, former staff director of the Joint Committee on Atomic Energy, accused Oppenheimer of treason. He wrote a letter to FBI Director J. Edgar Hoover claiming that "more probably than not" Oppenheimer had been a Russian spy from 1939 through 1942.

Fermi thought the charge absurd. He was outraged when it was announced that Oppenheimer, after years of service in which he had been in on the most important military secrets of the nation, was to have his security clearance suspended by the AEC while the charges were investigated.

Hearings were held by the Personnel Security Board of the AEC from April 12 to May 6, 1954. Although his illness was growing worse, Fermi testified on Oppenheimer's behalf on April 20. He pointed out that Oppenheimer "had rendered outstanding service during the war, that after the war his advice had been given after thorough study and in good faith. If it had not been taken, or if it was thought to be wrong, these facts offered no grounds for impugning his loyalty."

Fermi would have had much more to say in favor of Oppenheimer, but the review panel cut his testimony short in order to hear the next witness. In the climate of McCarthyism it probably would have made no difference if Fermi had his full say. Oppenheimer's motives for continued opposition to the H-bomb program were called into question. He "had exhibited weakness of character and had entered into association with unreliable characters." Oppenheimer was declared a security risk and his AEC clearance was permanently taken away from him.

Oppenheimer's reputation had been ruined. Fermi was shocked and dismayed. It was an additional burden to the illness which now held him in its grip.

On October 9, 1954, he entered Chicago's Billings Memorial Hospital for an exploratory operation. It revealed that Fermi had cancer of the stomach in a very advanced stage. Surgery to remove it was out of the question.

His old friend and associate Emilio Segre visited him there. Fermi's mind was still on the Oppenheimer case. He was particularly disturbed by the testimony of a colleague which he thought had been unethical. He told Segre he wished he had a chance to set this man straight. "What nobler thing for a dying man to do than to try to save a soul?" Fermi had smiled through his pain.

He was fifty-three years old. The cancer that was eating at him was the result of years of exposure to radiation. On November 29, 1954, the "Father of the Atomic Bomb" quietly passed away.

## CHAPTER 12 NOTES

p. 112    "it would be five to twenty years . . ." William Miller, *A New History of the United States,* p. 425.

p. 113    "One must take more time . . ." *Time Magazine,* November 25, 1946—"Education" section.

p. 113    "Fermi gave extremely lucid . . ." Emilio Segre, *Enrico Fermi, Physicist,* pp. 169–170.

p. 114    All quoted material on this page drawn from: *Pageant Magazine,* February, 1956, p. 36.

p. 116    All quoted material on this page drawn from: Richard Rhodes, *The Making of the Atomic Bomb,* p. 375 and p. 540 and p. 563.

p. 117    "The fact that no limits exist . . ." Emilio Segre, *Enrico Fermi, Physicist,* p. 165.

p. 117    "This explosion caused . . ." *Encyclopaedia Britannica,* Volume 2, 1970, p. 719.

p. 117    "three or four years . . ." William Miller, *A New History of the United States,* p. 425.

p. 117    "a huge magnet and a metal box . . ." Laura Fermi, *Atoms in the Family: My Life with Enrico Fermi,* p. 260.

pp. 122–123 All quoted material on these pages from: Emilio Segre, *Enrico Fermi, Physicist,* pp. 180, 181, 182.

# AFTERWORD
# THE LEGACY OF FERMI
# 1991

Enrico Fermi was one of the truly great scientists of the 20th century. Where some dealt exclusively with theory, Fermi put his theories to practical use. Where some changed the way we look at the world, Fermi changed the world itself the day he succeeded in setting off the first nuclear chain reaction under the bleachers of the football stadium at the University of Chicago.

A new source of energy had been tapped, and a new source of destruction. Fermi saw unlimited possibilities for using atomic energy to benefit mankind. But in the end he had to acknowledge the massive devastation he had helped to unleash, and this troubled him. It reminded him of something he had learned while still a boy in Rome, a principle which had guided him through his most complex experiments, a rule of thumb not just valid for physics but in all areas of life. It is simply this: For every answer, there is at least one more question.

Can an atom be split? When it is, what happens to its particles? Does splitting change them? Does the change release energy? How can that energy be harnessed? How can it be used to generate more energy? Are there side-effects to the release of this energy? Are the side-effects dangerous? Are there means of making them less dangerous? How much waste is created when nuclear energy is generated? What can be done with this waste? Is the disposal of nuclear waste more of a problem than the generation of energy is a benefit? Will the waste harm people more than the energy will help them? Will future scientists come up with answers to such questions, or will science flourish at the expense of humanity and its environment? Are the results of nuclear energy worth the risks?

If Fermi had lived, his attention surely would have turned to these questions. He might have answered some of them. He would surely have come up with other questions.

One of Fermi's great strengths was his ability to concentrate. Focus, young people are told, is a key to success in any field. Keep your eye on the ball and not on the bleachers.

The problem is that this can lead to tunnel vision. While watching the ball you can miss the play, the stolen base, the long slide, not to mention the storm clouds over the outfield. Great scientific discoveries often result from a narrow focus, but a narrow focus often hinders the scientist's full understanding of the effects his work will have.

Fire cooks meals, but it also burns down forests. Cars transport people, but their exhausts pollute the atmosphere. A nuclear chain reaction releases energy, but in a bomb it can kill on an unprecedented scale. There are often serious consequences to many scientific advances.

Fermi came to realize this and it became a cause of concern. Always his focus had been on science, but late in his life he saw the need to think in broader terms. He adopted a long-range global view. He set aside his long-held belief that nothing should stand in the way of scientific progress. He had the courage to change his mind.

Change is essential to science. To challenge the old beliefs and suggest new ones, to look at things in a different way—this was at the core of Fermi's genius. As a boy, with tops, he had shifted his focus from measurements to friction and its effects. As a young student he read works that challenged his teachers' theories rather than confirmed them. His pursuit of a "perfect" gas led him to the Fermi-Dirac Statistics. Measuring radioactivity, he noticed how measurements varied with the same element in different containers, and so changed the focus of his research, discovered the slow neutron, which earned him the Nobel Prize. Veering in another direction, he dared to dream of harnessing the energy of the split atom.

Always, his vision was spurred by a challenge. He *imagined* that if neutrons could be slowed down enough, a chain reaction would occur. Fermi *conceived* of an atomic pile, the first in history, and then he created it. He *envisioned* a mighty bomb to end the war, and only after that could it be built.

Early on in life, Fermi learned that there are many facets to science. If he started out with a problem in physics, it almost always lead him to mathematics. Chemistry offered a different perspective and fresh insights. Geology and astronomy both tied into atomic research. The deeper the question, the wider the net he had to cast to retrieve the answer.

Nor did Fermi hesitate to use his hands as well as his brain. When he needed a Geiger counter back in Rome, he made one himself. Two decades later he put together the Fermi Trolley. And when the first atomic pile was being built in Chicago, Fermi's hands were often black from piling its graphite bricks.

He came to appreciate that strands of knowledge intertwined not just in science but in life itself. Atomic research could not be separated from the world, from people and from politics. The Italian Fascists had taken credit for his discovery of the slow neutron. The United States military had taken over the chain reaction to develop the bombs that devastated Hiroshima and Nagasaki. Because of a government program, his patents had been given to American industry. The decisions concerning the use of his discoveries were taken out of Fermi's hands.

This did not make him bitter. He was not even sure it was a bad thing. In a democracy, perhaps the people—or their representatives—should decide something as important as how nuclear energy should be used. Truly, it was enough for Fermi that he had made such important contributions to the scientific knowledge of the world. Whether that knowledge is used for good or evil is—and will continue to be—a responsibility we all share.

# GLOSSARY

*alpha particles:* the positively charged nuclei of helium atoms.

*atom:* the tiny unit which, collectively, makes up all matter.

*atomic pile:* a crisscrossed structure of uranium and graphite, the size of which is determined by the point of which a chain reaction occurs.

*berylliosis:* a lung disease caused by inhaling beryllium.

*control rods:* cadmium sheets nailed to flat wood strips that were used to keep the layers of graphite and uranium separated in order to prevent premature fission in an atomic pile.

*chain reaction:* when the neutrons released by the splitting of an atomic nucleus in turn split other nuclei to produce energy in a continuous process.

*critical assembly:* an atomic pile carefully maintained at just below the point where a chain reaction can start.

*critical mass:* that point in the construction of an atomic pile at which a chain reaction is spontaneously triggered.

*cyclotron:* a device to energize atomic particles through magnetic action.

*ekarhenium:* the controversial short-lived "element" which appeared and then disappeared when Fermi bombarded U92 with neutrons.

*electron:* one of a group of tiny "planets" orbiting the nucleus of the atom.

*Fermi-Dirac Statistics:* formulas to determine how well various metals will conduct heat or electricity.

*fermions:* particles of gas that follow the laws laid down by the *Fermi-Dirac Statistics.*

*Fermi's Trolley:* a device made of lucite and mounted on wheels which can move elements within the chambers of the cyclotron.

*fission:* the splitting of uranium atoms into roughly two equal parts, a process releasing large amounts of nuclear energy; the first step in a chain reaction.

*Geiger counter:* a device which measures radiation.

*graphite:* a purer variety of the substance used in lead pencils that was used as a substitute for heavy water to filter and slow down bombarding neutrons in the atomic pile.

*gyroscope:* a wheel mounted in a ring so that its axis is free to turn in any direction while being observed.

*heavy water:* a manufactured substance in which the hydrogen molecule weighs twice as much as in ordinary water, making it an effective filter for slowing down bombarding neutrons.

*Irredentist movement:* an early 20th-century Italian nationalist group opposed to Austrian economic exploitation of Italy that counted future dictator Benito Mussolini among its members.

*Italian navigator:* code name for Fermi.

*jingoist:* expressing warlike bluster.

*Manhattan Project:* code name for the U.S. program to develop an atomic bomb.

*mare nostrum:* "our sea"—assertion of Italy's right to dominate the Mediterranean.

*neutron:* the unit of negative energy present in the nucleus of the atom.

*nuclear reactor:* an atomic pile designed to achieve a chain reaction.

*nucleus:* the center of the atom; the "sun" of its tiny "solar system."

*nutation:* the point at which friction causes a decrease in the speed of spinning so that a top starts to wobble.

*positrons:* positively charged particles released during radio-activity.

*precession:* the upward motion proceeding from the axis of a top when it is first spun that determines its vertical position.

*proton:* a unit of positive energy present in the nucleus of the atom.

*quanta:* units of energy; commonly used to refer to such units released by smashed atoms.

*Quantum theory:* the idea that light and energy are not continuous streams, but rather made up of many small units (called *quata*) in action.

*radiation:* energy released by smashed atoms.

*radioactivity:* another term for *radiation.*

*scaldino:* an iron pot in which students burned charcoal to keep themselves warm while studying.

*termodynamics:* science dealing with the relationship of heat and mechanical energy and the changing of one into the other.

*termonuclear fusion:* the extreme heat that melts down hydrogen into helium in an H-bomb.

*Triple Alliance:* a mutual defense pact between Germany, Austria and Italy from which Italy withdrew when World War I broke out.

*ZIP:* a safety lever in the Chicago atomic pile that was designed to stop a chain reaction.

# FURTHER READING

## Books about Enrico Fermi:

De Latil, Pierre. *Enrico Fermi, The Man and His Theories.* (Translated from the French by Len Ortzen.) New York, N.Y.: Paul S. Eriksson, Inc., 1966. This study of Fermi's life and work presents a balanced view of the controversy among the scientists involved in making and dropping the atom bombs.

Fermi, Laura. *Atoms in the Family: My Life with Enrico Fermi.* Chicago, Illinois: The University of Chicago Press, 1954. The human side of Fermi as a boy and man, husband, father, scientists, teacher and exile as seen through the eyes of the woman who shared his life.

Lichello, Robert. *Enrico Fermi: Father of the Atomic Bomb.* Charlottesville, N.Y.: Story House Corp. (SamHar Press), 1971. A simplified account, rich in anecdotes, of Fermi as student and innovator, Nobel Prize winner and "enemy alien."

Segre, Emilio. *Enrico Fermi, Physicist.* Chicago, Illinois: University of Chicago Press, 1970. The scientist as seen through the eyes of a close associate who was privy to both the ongoing development of Fermi's theories and to the questions of conscience raised by his work.

## About Fermi's Colleagues:

Clark, Ronald. W. *Einstein: The Life and Times.* New York, N.Y. & Cleveland, Ohio: World Publishing (Times Mirror), 1971. A definitive, full-length biography of the "father" of quantum theory.

Lawren, William. *The General and the Bomb: A Biography of General Leslie R. Groves, Director of the Manhattan Project.* New York, N.Y.: Dodd, Mead, 1988. A portrait of the complex military careerist who was in charge of the scientists at Los Alamos.

Stern, Philip M. *The Oppenheimer Case: Security on Trial.* London, England: Hart-Davis, 1969. An even-handed and all-encompassing account of the controversial removal of J. Robert Oppenheimer's security clearance by the Atomic Energy Commission.

York, Herbert Frank. *The Advisors: Oppenheimer, Teller and the Superbomb.* San Francisco, California: W. H. Freeman, 1976. How all of the top atomic scientists, including Fermi, lined up on the questions surrounding the development and testing above ground and underground of the hydrogen bomb.

## About Science and Nuclear Physics:

Crabtree, Harold. *An Elementary Treatment of the Theory of Spinning Tops and the Gyroscopic Motion.* New York, N.Y.: Chelsea Publishers, 1967. An explanation of such determinants of object motion as friction, precession, nutation, etc.

Hubbard, L. Ron. *All About Radiation.* Los Angeles, California: Scientology Publications Organization, 1957. Hubbard's clear explanation of what radiation is, its containment and the dangers of its release, etc. is made easy to follow by charts and drawings based on a theory of atomic structure that remains valid today.

Macauley, David. *The Way Things Work.* Boston, Massachusetts: Houghton-Mifflin, 1988. Contains an easily understood section on the way nuclear reactors work.

Solomon, Joan. *The Structure of Matter.* New York, N.Y.: John Wiley & Sons, 1974. Traces the evolution of molecular theory from myth through early civilizations on up to the first theoretical work on atoms leading to the atomic and hydrogen bombs.

Walter, Jearl. *The Flying Circus of Physics with Answers.* New York, N.Y.: Wiley, 1977. Observations on the behavior and reactions of objects in a variety of circumstances from which general theorems of physics may be drawn.

## About the Manhattan Project, Los Alamos and the Dropping of the A-Bomb:

Groueff, Stephane. *Manhattan Project: The Untold Story of the Making of the Atomic Bomb.* Boston, Massachusetts: Little

Brown and Company, 1967. A short, fast-moving narrative of the development of the bomb from initial research at Columbia University to Hiroshima.

Hersey, John. *Hiroshima.* New York, N.Y.: The Limited Editions Club, 1983. Eyewitness accounts of the immediate devastation following the dropping of the bomb.

Mee, Charles L., Jr. *Meeting at Potsdam.* New York, N.Y.: M. Evans & Company, Inc., 1975. Based on declassified government documents and personal memoirs, this reconstruction of the 1945 meeting between Truman, Churchill and Stalin puts into perspective the political considerations behind the decision to drop the bomb.

Rhodes, Richard. *The Making of the Atomic Bomb.* New York, N.Y.: Simon & Schuster, 1986. The definitive work on the subject, this award-winning account is long, but both thrilling and readable.

## About Italy and Mussolini:

Kennan, George Frost. *The Fateful Alliance: France, Russia and the Coming of the First World War.* New York, N.Y.: Pantheon Books, 1984. Italy's dependence on Austria and eventual siding with the Allies, as well as Mussolini's swing from socialist pacifism to jingoism, weaves in and out of this account of Europe's shifting balance of power.

Lyttle, Richard B. *Il Duce: The Rise and Fall of Benito Mussolini.* New York, N.Y.: Atheneum, 1987. A considered reappraisal of the Italian dictator's evolution from socialism to fascism and from popular conqueror to assassinated despot.

Michaelis, Meir. *Mussolini and the Jews: German-Italian Relations and the Jewish Question in Italy, 1922–1945.* New York, N.Y.: Clarendon Press, 1978. This study places Fascist anti-Semitism in the context of nationalist xenophobia and demonstrates how Mussolini's measures against Jews were frequently thwarted by the Italian People.

Mussolini, Benito. *My Autobiography.* New York, N.Y.: C. Scribner Sons, 1928. A self-serving account of his early years and rise to power.

# INDEX

Italic numbers indicate illustrations

| DATE | | | |
|---|---|---|---|
| | | | |
| | | | |
| | | | |
| | | | |
| | | | |
| | | | |
| | | | |
| | | | |
| | | | |